Journalism and the Metaverse

Journalism and the Metaverse

John V. Pavlik

ANTHEM PRESS

Anthem Press
An imprint of Wimbledon Publishing Company
www.anthempress.com

This edition first published in UK and USA 2024
by ANTHEM PRESS
75–76 Blackfriars Road, London SE1 8HA, UK
or PO Box 9779, London SW19 7ZG, UK
and
244 Madison Ave #116, New York, NY 10016, USA

© John V. Pavlik 2024

The author asserts the moral right to be identified as the author of this work.

British Library Cataloguing-in-Publication Data
A catalogue record for this book is available from the British Library.

Library of Congress Cataloging-in-Publication Data
A catalog record for this book has been requested.
2024931898

ISBN-13: 978-1-83999-012-0 (Hbk)
ISBN-10: 1-83999-012-0 (Hbk)

This title is also available as an e-book.

CONTENTS

Introduction   1

1. Disrupting Journalism: Setting the Parameters of the Metaverse   3

2. Principles of Journalism in a Virtual World: Ensuring
    the Pursuit of Truth   25

3. Structural and Systemic Considerations in the Metaverse:
    Regulations, Economics, and Newsroom Organization   39

4. Metaverse Reporting: Looking Inside and Out   63

5. Journalists as Avatar: Evolving toward a Hybrid Model
    of Human and AI Reporters   77

6. Users of the Metaverse: A Virtual Public   95

7. Research on XR-Based Journalism: Designing News
    for the Metaverse   111

8. Concluding Reflections: Achieving Journalistic Excellence
    in the Metaverse   131

References   143

Index   175

INTRODUCTION

Journalism has been in a near-constant state of tumultuous change since the advent of the Internet and other digital technologies. The consequences of digital disruption have been profound for journalism and society. Newspapers have rapidly vanished from the landscape as the long-standing economic underpinnings of commercial news media have shifted in seismic fashion away from the analog world of media and into the digital realm of modern communication platforms. Citizens and reporters alike increasingly use their mobile and other digital devices to access news and other media content as well as create it. Data in fact show that three-quarters of the world's population access the Internet solely via their smartphone (Handley, 2019). They do so in almost nonstop fashion fueled by the increasing ubiquity of wearable devices such as smartwatches and other smart media continuously connected to the Internet. As augmented reality and virtual reality gradually build momentum in the marketplace, the Metaverse is likely to further fuel the disruption of journalism. Analysts point to the media as among the five industries the Metaverse is most likely to transform (Danise, 2022; Raftery, 2022; Ball, 2022).

The thesis of this book is that the Metaverse presents a two-part challenge to journalism. First, the Metaverse is a new arena for news reporting and journalism needs to establish a presence in this arena to cover important stories and reach a growing audience especially younger audiences engaging in the Metaverse. Second, the Metaverse signals an opportunity for news media innovation in storytelling and other journalism methods of production, distribution, and management. Failure to meet these challenges will likely lead to a lost opportunity for journalism to become a vital platform in the Metaverse as it emerges in the 21st century. Moreover, journalism as an institution and public good will likely further erode as it has done in the age of the Internet to date.

In his acclaimed novel *Snow Crash*, science fiction writer Neal Stephenson in 1992 introduced or at least popularized the term Metaverse as a virtual world where humans might one day all live, work, and play. In this virtual realm, there also might be journalism to create and experience. Stephenson's was a dystopian vision, a cautionary tale. He wrote, "Hiro [the protagonist] is in a computer-generated universe that his computer is drawing onto his goggles and pumping into his earphones [...] this imaginary place is known as the Metaverse" (p. 24). The word Metaverse itself is a portmanteau, combining meta, or over or beyond, and universe, thus implying that the Metaverse subsumes all life or worlds. It is worth noting that the term Metaverse long predates Stephenson's usage, with Metaverse being used in the 19th century in reference to poetry about poetry.

Chapter 1

DISRUPTING JOURNALISM: SETTING THE PARAMETERS OF THE METAVERSE

With the COVID-19 pandemic pushing humans to live increasing portions of their lives online, the Metaverse has gained traction, and public and corporate interest in it has accelerated. A 2022 survey by Gartner (IANS, 2022) shows that nearly two-thirds (58%) of the public has at least heard of the Metaverse. But despite having heard of the Metaverse, almost everyone is still wondering just what it is, or will be, and what its implications are for journalism, media, and society in a post-pandemic world; many teens don't care about the Metaverse despite interest in virtual reality (VR) (Steele, 2022; Tran, 2022; Milano, 2022). The Gartner survey shows that 94% of the public would be uncomfortable trying to explain the Metaverse to someone else. A survey by Global consulting giant Accenture shows that 71% of business executives think the Metaverse will change how businesses operate and interact with their customers (Torkington, 2022). Examining the nature of the Metaverse and outlining the scope of its implications for journalism are the topics of this chapter.

Immersive and Interactive Internet

As of this writing, there is no single, agreed-upon definition of the Metaverse and how it will be designed and built (Takahashi, 2022; Green & Works, 2022). For the purpose of this book, the Metaverse is defined as an immersive and interactive virtual world or constellation of networked virtual worlds. As such, the Metaverse is a key development in the next generation of the broadband, publicly accessible Internet. Although anyone may have access to the Metaverse as they might enter a privately owned shopping mall in the real world, that access may be contingent on a variety of conditions, from affirming acceptance of a user agreement to paying for entry or the requisite technological devices. Immersive means it is an enveloping, 3D form of media and communication. Immersive media today come most typically in the form of augmented reality (AR) and VR, which are labeled eXtended Reality (XR). These forms are three dimensional (3D); they have depth. Although they are virtual, they often are digital analogs to the physical world which similarly is enveloping. Immersive media not only have depth but they also surround the user in a 360-degree virtual space. And in contrast to traditional media, users are active, rather than passive, in their content experience. Users can look about in any direction: forward–backward, up–down, left–right. This gives the user three degrees of freedom (DoF) in that virtual space. Immersive media that also support user movement in a 3D, 360-degree space give the user six DoF. That is, the user can not only look but also move in any direction, left–right, forward–backward, or up–down, just as they might in the physical world. These six DoF are essential qualities of flight simulators, for example, which media historians have noted were early 20th-century precursors to the VR platforms of the 21st century, and likewise the Metaverse. The Link Trainer, for instance, was an electromechanical flight simulator Edward Link invented in 1929 and patented in 1931 (Internet Society, 2017). For journalism, DoF represent the potential transformation of news storytelling in which users not only read, see, or hear a story; they enter into it as a participant or witness to news events or other subject matter. Illustrative of the type of storytelling possible in an immersive environment is "The Infinite" VR. A collaboration between NASA and Felix & Paul Studios, a Montreal-based VR production company, the experience takes the user on an immersive journey aboard the International Space Station (ISS). The VR experience blends more than 200 hours of VR recordings on the ISS captured over about three years. It is offered as an in-person experience. But in the future, in a mixed reality (MR) Metaverse, such an immersive experience becomes possible for innovative work in journalism on other nonfiction reporting. MR utilizes spatial computing, which involves the creation of a 3D environment generally across all modalities and enables layering interactive digital content into any space. Although the Metaverse itself is still in an unsettled state, industry forecasts suggest that among the most likely jobs to exist in the next decade is "Metaverse storyteller" (.cult, 2022). This presents an interesting career possibility for journalists and anyone seeking a career in journalism.

Interactive means both user-to-user engagement and an exchange between the user and the enveloping content experience of a virtual world, especially enabled by artificial intelligence (AI). Broadband refers to the high-bandwidth or high-speed nature of Internet connectivity that the Metaverse will require. These elements are also impactful for journalism. Traditional journalism has treated the audience largely as passive recipients of the news. In the interactive, broadband Metaverse environment, journalists should reimagine the user as active, not only consuming the news but commenting on it, and sometimes contributing to it or even serving as a co-creator of journalism, or virtual citizen journalist, typically subject to fact-checking by crowdsourcing.

The Immersive User Experience: Multisensory Envelopment and Engagement

Essential to the viability and success of the Metaverse is the user experience (UX). Immersive media feature enveloping sound, and potentially engage all the human senses, including touch or haptics, taste and smell, or even digital-only senses, such as those that might involve data synthesis or brain–computer interfaces (BCI). Commercial BCI systems are in development, including by Elon Musk's company Neuralink and Paradromics Inc. (Q.ai, 2022; Gilbert, 2023). Coincidentally, the source cited for this Neuralink news is Q.ai, an AI-powered investing app from Forbes that also writes news stories (Blake, 2022). BCI enable two-way communication between human and machine through thought and electrical impulses via a microchip implanted in or otherwise connected (possibly wirelessly) to the user's brain. The Neuralink system is intended initially for persons with disabilities, but later generations could be designed for any interested user and could enable a Metaverse persistent and almost inescapable. BCI systems such as Neuralink may bring society disturbingly close to entering the human–machine fusion as depicted in the Hollywood blockbuster series "The Matrix" or toward what futurists call "the singularity" (Jones, 2006). The Food and Drug Administration has raised questions about BCI, however, and unless its concerns including ethics and security can be resolved BCI may not develop, at least in the United States.

For journalism in the Metaverse, BCI means news or news-related experiences that the user could directly engage virtually instantly through thought. One day, news could be delivered directly into the minds of the public or at least those of their Metaverse avatars (thus, an avatar might learn the news before its human counterpart). There is a digital precedent. In 2012, Datamir was among the first firms to develop and employ algorithms to digest, synthesize, and publish news. Datamir's algorithm monitored online sources, especially social media, and translated posts into actionable trading information. This news, however, was not delivered to a human audience. Instead, it was delivered electronically to other online algorithms encoded to make trades in the market when certain predetermined conditions were met. This ushered in what is called news-based trading. Gupta (2015) states that this type of automation "helps report the latest business news up to 54 minutes faster than conventional news coverage." It is a key part of the advent of algorithmic trading. It also suggests that in a digital, virtual world, avatar journalists may produce and share news with other robotic avatars populating the Metaverse.

360-Degree Envelopment, 2D and 3D Experience

Some versions of the Metaverse support users accessing the virtual environment from a computer or handheld digital device (e.g., smartphone or tablet). These experiences may not be fully immersive, or enveloping, in a 3D and 360-degree fashion for the user. Rather, they are typically more of a 2D experience. Although these virtual worlds may feature a 3D design, the user is not sensorially fully enveloped visually, aurally, or otherwise, although new holographic displays in development may change this. Journalism in these 2D environments will be more conventional in design.

Yet 2D interfaces are advantageous for several reasons. One is that a wider swath of the global public can access such a form of the Metaverse. Nearly 5 billion persons worldwide have a digital device, a smartphone, internet-connected tablet, or computer (Statista, January 2022). Thus, these billions can immediately enter a 2D Metaverse, once they have installed the requisite app. By contrast, about 200 million persons have a VR headset as of early 2023, making the human population that could enter the Metaverse dramatically smaller. For journalism, the benefits of reaching billions in a 2D Metaverse would be substantial. These benefits include potentially informing more people and building a larger user base to sustain journalism economically.

The Pre-Metaverse

Contemporary 2D forms of a nearly immersive Internet constitute the pre-Metaverse. They are on the path to the fully immersive and interactive Internet or Metaverse. In this sense, Zoom or other video conferencing platforms are early social forays into a limited form of the pre-Metaverse. Journalism can utilize these 2D media forms in the near term to gain an inexpensive though limited stake in the Metaverse.

Examples of existing 2D Metaverse variants include a variety of commercial platforms. *Second Life* is among the most well known and influential. Linden Lab developed *Second Life*, launching the platform in 2003. *Second Life* provides the basic template for many corporate visions of the Metaverse (Virgilio, 2022). A quick glance at the *Second Life* platform makes clear that it is essentially a direct precursor to Meta's concept of the Metaverse and likely was its inspiration in large part at least in terms of design (Au, 2008). *Second Life* itself derives from earlier visually based virtual worlds such as *The Sims*, launched in 2000 (Virgilio, 2022).

Second Life is an online, animated, computer-generated simulation, an artificial reality that mimics, or parallels, the physical world. It is an alternative digital world that users design and populate. *Second Life* features user-customized avatars that represent the user within its virtual world. Users access and experience *Second Life* via a computer or mobile digital device and view the world from a third-person perspective. The viewpoint is as if a camera on a drone hovers over the user's avatar, sees the avatar and what the avatar sees, interacts with others, or hears sounds, including ambient acoustical effects such as birds chirping when outdoors. The user controls what their avatar does and looks like within *Second Life* as well as how their avatar interacts with other avatars or virtual objects. The content of *Second Life* is user-generated and the platform has about 200,000 active daily users worldwide as of 2022 (Virgilio, 2022). *Second Life* has its own currency and features a variety of qualities or dimensions that parallel the real world, a sort of digital twin to the physical world. Users in the form of avatars can meet other residents (or users) of *Second Life*, and build, create, shop, and trade virtual property and other things. During the pandemic, *Second Life* claimed a GDP of about $650 million (Virgilio, 2022).

Professors have held courses in *Second Life*. Artists have created digital art in *Second Life*, and musicians have performed virtual concerts attended by other residents of *Second Life*. Journalists have produced news reports within and about *Second Life* (Brennen and dela Cerna, 2010). Virtually anything possible in the physical world can be reproduced digitally in *Second Life*. Consequently, it is intended for persons aged 16 or older, although 13–15-year-olds are allowed within a part of *Second Life* managed by a sponsoring institution such as a school. *Second Life* is in many ways modeled after much older virtual worlds called massively multiplayer online role-playing games (MMORPG), which were themselves derived from the first multiuser games developed in the mid-1970s shortly after the invention of the Internet.

Video Games

Video games constitute a huge industry and global social phenomenon, in terms of both revenues generated and usage worldwide. Video games are no doubt another factor contributing to the corporate interest in designing the Metaverse. Statista reports that in the United States alone the value of the video game industry is more than $97.67 billion (Clement, 2022). US consumers spent $11.6 billion on video games in 2020. Market leaders include Nintendo, the Japanese gaming giant and console maker. Given the capacity of the Metaverse as a gaming platform, it is not surprising Nintendo President Shuntaro Furukawa has said, "The metaverse is attracting the attention of many companies around the world, and we believe that it has great potential," per an English translation from *VGC* (Hayward, 2022). More than 2.5 billion people play video games worldwide (Yanev, 2022). At least 60% of Americans play video games daily, and nearly half (45%) of gamers are women. Many video games are played online and in concert with others and are a possible pathway to the Metaverse.

These numbers increasingly exceed or even dwarf the size and engagement of the public with regard to mainstream news media. Traditional news media such as newspapers are rapidly shrinking in terms of public engagement, as the vast majority of the public turns to online or digital platforms (e.g., mobile devices, smart TVs, podcasts) to get news. For adults aged 18–29 years old (especially Gen Z), newspapers are the last news source turned to. Moreover, most prefer free news sources. Worldwide, 70% say they never turn to a paid print or digital news subscription service. About half (48%) of the US public reports getting news from social media on a regular basis (Watson, 2022). Revenues for traditional news media are low compared to the realm of video games. Revenues for the three major cable news channels in the United States totaled $6.7 billion in 2020: $1.7 billion for CNN, $2.9 billion for Fox News, and $1.1 billion for MSNBC per Kagan, a media research group in S&P Global Market Intelligence. These represented a slight annual revenue increase (3–5%). License (affiliate) fees declined about 1% overall while advertising revenue grew 9–12% (Pew Research, 2021). Revenues for the newspaper publishing industry in the United States were approximately $26.5 billion in 2019. Advertising generated about two-thirds ($15 billion) and much of the rest came from subscription or other user fees. Total newspaper revenues are forecast to decline by $21 billion by the end of 2024 (Watson, 2022). The situation for broadcast news engagement and revenues is mixed although somewhat more encouraging. Perhaps at least partly due to the pandemic and people staying home more, local news audiences grew for the ABC, CBS, NBC, and Fox affiliates in 2020 in the United States. Evening and late-night newscast audiences were up by 4% per Comscore data cited by Pew (2021). The midday news slot grew by 10% in 2020 while morning news audiences dropped by 4%. Despite these trends, overall audiences for local news are in long-term decline.

The evening local TV news audience totaled 3.73 million in 2020, down from 4.10 million in 2016. Audiences for late-night and morning local TV news similarly have declined in the past decade. Parsing out the revenues specifically for local TV news is difficult. But the overall revenues for local TV over-the-air advertising totaled $18.4 billion in 2020, up 8% from 2019 (Pew Research Center analysis of MEDIA Access Pro & BIA Advisory Services data, 2020). Some portion of that is attributable to news programming, but absent the data it is hard to more than speculate. Pew adds that "revenue for the 839 local TV stations defined as 'news-producing stations' (stations that have a news director and are viable, commercial and are English-language affiliates in the U.S.) was $15.3 billion, according to the BIA Advisory Services database." The audience for all the three network nightly newscasts grew significantly in 2020, even more when streaming is factored in (Barthel, 2021). Per Comscore TV Essentials data cited by Pew, "ABC evening news viewership grew 16% to 7.6 million viewers in 2020, following an 11% increase in 2019. CBS evening news viewership grew 7% to about 5 million viewers in 2020, while NBC viewership rose 8% to 6.5 million." Kantar data cited by Pew indicate ABC saw a 22% increase in ad revenue in 2020, while CBS and NBC each saw an 8% increase (Winslow, 2021). Radio news is important but relatively small in terms of overall audience and revenues, and both are in long-term decline, especially terrestrial over the air. Satellite and online news delivery and podcasting are growing in audience.

The situation for journalism is troubling, especially since newspapers have historically been the primary source of original reporting for the total news ecosystem. Something fundamental needs to change or there will be little left to mainstream news media, at least in the United States, beyond a handful of major national or international news organizations such as *The New York Times*, cable, and the broadcast networks, although these are rapidly shifting toward digital platforms and streaming news providers. The Metaverse points to a possible pathway to a future in which journalism plays a major and robust role. The design of news in the Metaverse is also a key consideration. The following popular video games from the pre-Metaverse may point a way for journalism to adapt.

Fortnite is a hugely popular multiplayer role-playing online animated video game and it suggests a direction in the design of the Metaverse (Sullivan, 2022). Epic Games designed and launched *Fortnite* in 2017. *Fortnite* has a reported 350 million users and 15 million simultaneous players at any moment worldwide (Garton, 2022), so it is several orders of magnitude larger than *Second Life*, at least from a user point of view. It features a 3D-designed virtual environment in which users participate via customizable avatars. As with *Second Life*, users participate in *Fortnite* via computer or handheld device and can download a free app to play. Users can create worlds and battle arenas and, as with *Second Life*, make purchases, but with real money on in-game purchases, and this is how the virtual world generates its substantial revenues for its real-world parent company. *Fortnite* generates hundreds of millions of dollars in annual user revenues and generated a reported $9 billion in gross revenue in its first two years (Clark, 2021). As of April 2022, Epic Games was valued at $32 billion (Wile, 2022). The monetary success of *Fortnite* likely also plays a role in inspiring other corporate forays into the Metaverse. However, *Fortnite* has come under regulatory scrutiny for its business practices which apparently include ethically as well as legally dubious actions. In December 2022, the US Federal Trade Commission (FTC) fined Epic Games $520 million for practices that both compromised the privacy of child users and mislead users into making unwanted in-game purchases (Wile, 2022). The ethically and legally questionable practices of Epic Games suggests that mainstream news media have one major advantage in the arena of games in the Metaverse. News media historically have placed a premium on ethical practices, although with the rise of hedge fund ownership of news media this historical pattern may start to erode as the priority is increasingly placed on profits and not necessarily news quality. Still, if news media can maintain their commitment to the highest ethical standards and introduce an approach to news delivery that increasingly features game design (e.g., immersive news games), news media might grow in prominence and user engagement in the Metaverse, and potentially see their revenues rebound.

Another popular video game platform is Nintendo Switch *Animal Crossing*, typically played online and with other game players. Statista (9 February 2022) reports lifetime sales of *Animal Crossing: New Horizons* are 37.62 million units as of early 2022. In *Animal Crossing*, players create their own customized islands with a variety of digital wildlife and invite others to join them on their creative isles. In this sense, *Animal Crossing* is another example of a pre-Metaverse platform. This platform-specific game became hugely popular during the COVID-19 pandemic, as people stayed home and often sought pleasant but stimulating digital diversions with social interaction. It is possible that such pre- or full Metaverse experiences could prove popular in a post-pandemic real world. Much of the success of *Animal Crossing* derives from the intersection of enabling user creativity, community building, and engagement. For journalism, a similar blended strategy for news games in the Metaverse may prove pivotal. News games that highlight interactivity and community along with encouraging users to contribute to those games and not just play them may be particularly appealing to younger audiences. Younger audiences have shown an increasing disdain for traditional news media while being drawn increasingly to social and interactive digital media that enable their own creative expression or contribution.

Tapping VR, AR, and MR

Most contemporary visions of the Metaverse delve into the realm of VR. VR-based Metaverse conceptualizations require users to don a headworn display, a VR, AR, or MR headset. Future generations of the Metaverse are likely to extend well beyond VR-only and into AR and MR, which comprise much of the overall realm of XR, including WebXR via laptop computers. The significance of this potential development is that entering the Metaverse may become something users can do anytime anywhere and without necessarily the encumbrance of a headset or other wearable. Via AR and MR, digital content and interactivity are layered on the user's real-world experience. Utilizing an AR or MR design, the Metaverse could enable users to enter and engage in interactive and immersive experiences at any time or at any location while out and about in the real, physical world. In other words, the digital twin can be layered into the physical world in a potentially ubiquitous fashion. Just as tools like Google Maps are available for mobile users, a mobile Metaverse could transform users' daily lives. For journalism, news could be available at any time or place and engaged immersively and interactively.

Presence

Through sensory envelopment, immersive media such as VR also can create a state of psychological immersion. That is, immersive media can create for the user the mental state of being, or feeling, as if they actually are inside, or present, in a virtual environment. This effect is called presence. Research suggests that presence is one of the most important factors that define and shape UX in virtual environments and influence a host of outcomes of that experience. Biocca and Delaney (1995) were among the first scholars to examine the role that immersive media can play in generating a sense of presence, or telepresence, for the user. Presence, or tele-presence, refers to the user's experience at a distance or virtual presence. Subsequent research has indicated that generating a sense of presence can be essential in shaping UX in various forms of immersive content, including immersive journalistic content (Kim and Biocca, 2006). Using an experimental research design, Archer and Finger (2018) found that 360-degree video news stories can produce a greater sense of presence than when those same stories are told using conventional narrative approaches, particularly text, photos, and linear, standard field-of-view video, and generate greater empathy among those who experience that immersive news content. Sundar, Kang, and Oprean (2017) at the Media Effects Lab at Penn State University likewise have found experimentally that 360-degree video for news stories generates a greater sense of presence for news stories and also can facilitate greater empathy and other learning outcomes such as recall. Vettehen, Wiltink, Huiskamp, Schaap, and Ketelaar (2019) have found that users evaluate 360-degree video news as more enjoyable and credible than 2D news and recognize and better understand 360-degree video than 2D news. However, Sundar et al. (2017) caution that journalists should use immersive experiences with caution or risk eroding the credibility of their news stories if immersive approaches are used as a gimmick. In the Metaverse, news media have an opportunity to develop immersive, enveloping news content that likely will have a receptive audience that is already engaging in VR. One of the key considerations for news media is the extent or portion of the overall public who will be participating in an immersive and enveloping Metaverse in the near term.

Research shows that other forms of immersive content also generate presence (Miller & Bailenson, 2021). Further research is needed to evaluate the impact of emerging immersive media forms such as cinematic VR on user presence; a fifth of those surveyed in 2022 say they are interested in experiencing cinema inside the Metaverse (Tran, 2022). In this spirit, documentary VR approaches may play a key role in Metaverse journalism and merit scholarly investigation, especially per their ability to generate presence and related outcomes such as empathy.

Presence is a multidimensional concept. Among the dimensions most relevant to VR experience is "presence as transportation." Slater and Usoh (1993) state that this "conceptualization captures the sense of presence in which media users feel that they or other objects, people, or environments have been transported [...] and leads to the 'suspension of disbelief that they are in a world other than where their real bodies are located'" (7, p. 222). For journalism, this suggests the potential to put users into the role of an eyewitness to news events and scenes and highlight their sense of presence in those news spaces. For the user, this may be a powerful new way to experience the news, potentially increasing understanding and empathy. Perhaps even more importantly for the future of the Metaverse and journalism, Vettehen et al. found that presence mediates 360-degree news viewing enjoyment and credibility. Since news experience and engagement are something the public does as a choice and not a requirement, ensuring that the news experience in the Metaverse is something people enjoy will be essential or the audience for MV journalism will not develop. Likewise, empathy and other factors related to immersive media experiences also are multidimensional. For journalism, this implies that increasing empathy may take different forms, such as seeing something from another perspective or understanding an issue from different viewpoints, and news content should be intentionally designed accordingly.

User Perspective

In this context, the user point of view or perspective is essential to the immersive news experience. Much typical media content has tended to employ a third-person perspective in which the user observes, reads, or hears the story or events as if an outside observer, listener, or reader. Some documentary productions have featured a first-person perspective. In this perspective, the viewpoint of the viewer is as if they are present on the scene as an actor or participant in the situation depicted on screen or via audio. In fact, old-time radio news plays such as CBS's "You Are There" from the 1940s placed the user inside the story (e.g., "You Are There: The Assassination of Abraham Lincoln," 7 July 1947, Internet Archive, 2023).

Heckman (2021) offers a practical definition of the first-person perspective. "A first-person point of view is a character perspective that's used to relay the thoughts and feelings of a character or entity within a story. [...] In video games and movies, the term 'first-person point of view' is used in reference to the perspective of the camera. For instance, the 'first-person shooter' is a video game genre in which the player controls a character (shooter) from their vantage point." Research suggests that in immersive media, the use of a first-person perspective affects the user sense of presence, and thereby can affect outcomes such as empathy. Some popular simulation games allow users first-person engagement in mundane behaviors virtually, such as power washing a wall or preparing a meal. News simulations that allow users to first-person role-play in a news event recreation could prove highly engaging.

Immersive media increasingly feature a first-person perspective. In contrast to the older *Second Life's* third-person viewpoint, the newer *Fortnite* offers users a first-person mode if desired. VR experiences designed for the Meta Quest (3 as of 2023) or other contemporary VR platforms likewise often feature a first-person perspective. Moreover, in 2021, the Quest introduced advanced haptic controller features for the user which, combined with head and hands tracking, enable the user to see their hands from the vantage of their own eyes in the virtual environment and feel objects virtually as if they are touching them. Researchers call this a sense of virtual "embodiment" in which the user gains the sense their own being is part of a virtual space (Kilteni, Groten & Slater 2012; Ahn, Bailenson & Park, 2014). Embodiment generates for the user a heightened sense of presence by delivering a virtual experience more closely approximating their experience in the physical world. This could be an opportunity for Metaverse journalism to generate pro-social outcomes as a consequence of diverse and inclusive immersive news content based on news experiences featuring a haptic first-person perspective.

Although not exactly pro-social, MVR has designed a haptic experience that allows the user to virtually experience the famous Oscars Will Smith slap (Melnick, 2022).

Spatial Sound

Other features or affordances of virtual environments also shape the UX and thereby may impact presence (Gibson, 1966, 1977; Flyverbom, Leonardi, Stohl & Stohl, 2016). Among the most important are the qualities of the acoustical or sound space. Spatial sound presents the listener with an audio experience that maintains the 3D and 360-degree integrity of the overall sound space (Tashev, 2019). Spatial sound is increasingly the standard for immersive environments. It goes beyond stereophonic audio and is important to increasing the user's sense of presence. Spatial sound maintains for the user the integrity of the audio elements of the virtual experience even if the sound source or the user moves about visually or in an implied fashion in the space. In other words, spatial sound maintains the locational accuracy or integrity of that sound space relative to the user. If a sound source moves closer to or farther from the user, or behind or above the user, the audio experience adjusts in a corresponding fashion so that the user perceives it as coming from the sound source in its implied location, even as it moves about. This is a vital feature in generating the verisimilitude of narrative experiences in virtual environments (e.g., in cinematic VR or documentary VR), virtual social interactions that involve audio, and any other virtual experience that includes sound. Spatial audio, therefore, is a key element to making a virtual news experience seem real.

Anyone who has ever experienced a compelling VR experience likely knows how real it can feel when sight, sound, and potentially haptic or tactile feedback are delivered in authentic fashion. A sense of presence can be very powerful and convincing in VR. Research has shown how impactful well-designed VR experiences can be, especially in Virtual Reality Exposure Therapy (VRET), which can play a key role in treating mental health problems including anxiety and depression (Lanier, 2022; Lindner et al., 2019). VR can be so powerful that for the user it can be overwhelming or nearly so, especially emotionally. Even riding a VR rollercoaster can take on a sense of reality, generating intense emotional response such as the thrill of a rapid descent. As authenticity and truthfulness are fundamental values in journalism, the potential of haptic experience may prove vital to journalism in the Metaverse and should be used with care.

In 2020, as an attendee at the online Tribeca Film Festival, the author experienced the emotional intensity of a cinematic VR representation

Figure 1.1 Saturnism, a cinematic VR experience.

of Spanish painter Francisco Goya's 19th-century masterpiece *Saturn Devouring His Son* (Pavlik, 2021). In this experience titled *Saturnism*, the user is inside a dark cave (Figure 1.1) (Cultural Services, 2020). Gazing about the faintly illuminated 360-degree space, the user hears a strange crunching sound. The user only gradually comes to realize the giant God Saturn is standing nearby, devouring his son. The user gets the distinct impression they could be the next course in Saturn's meal. The cinematic VR experience was brief, just about six minutes, but even that was long enough for it to be unsettling. Adding a haptic dimension to this experience might have made it emotionally overwhelming.

Goya's "Saturn" and its VR representation is something of a metaphor for where society is collectively headed in the digital age. In terms of the Metaverse, the metaphorical message might be an allegory about the potentially harmful consequences an immersive and interactive Internet might bring, that is, Saturn devouring his son in VR is a cautionary tale of humans and technology. Still, more cultural events including film festivals are moving online and into the Metaverse. In 2022, the Sundance Film Festival took place in the Metaverse. One attendee who gave the experience a positive review affirmed its capacity to generate presence, stating, "Attending the festival in virtual reality was surprisingly similar to being there in the real world" (Zeitchik, 2022).

Unintended Consequences

The Metaverse offers potential for innovation, virtual presence, and revenue generation in the digital realm. Certainly, business leaders such as Zuckerberg, Nadella, and Musk may see it as such or their billion-dollar investments would make little sense (Wayt, 2021). It is not clear how many journalism leaders see potential in the Metaverse.

But the unexpected impact of the Metaverse may be much more profound, especially in the long run. When Vinton Cerf and Robert Kahn envisioned the Internet and developed The Transmission Control Protocol (TCP) and the Internet Protocol (IP) (TCP-IP), which are the basic technical rules that make the Internet possible, they were working for the United States Department of Defense (TAL Technologies, 2022). Cerf and Kahn intended the network of networks, or Internet, as a means to enable communications among computers that could be located anywhere in the world (or even beyond into outer space), tolerant of faults, infinitely expandable, and would support new and unforeseen applications, as long as they conformed to TCP-IP.

But Cerf and Kahn may not have imagined how the Internet would lead to the global disruption of journalism. They may not have anticipated the global and rapid spread of mis and disinformation about phenomena such as the COVID-19 pandemic and its associated vaccines. Cerf and Kahn likely could not have predicted the destructive impact social media could have on democracy and democratic processes. These are among the unintended consequences of the Internet.

Sociologist Robert K. Merton in 1936 articulated the concept of the unintended consequences of social action. He argued that oftentimes the most significant and long-lasting impact of social action, including new technologies, is not the intended purpose for which that action or technology was ostensibly designed. Rather, it is the unintended impact that is more far reaching, enduring, and potentially devastating. As such, among the first things any innovator or inventor should consider is what unintended impact their creation might generate, for society, the world in general, or even beyond the planet. This is just as true for journalism as any other endeavor.

Whether innovators of 19th-century newspaper publishing or 20th-century broadcasting envisioned, the long-term implications of their innovations are unclear. Journalism pioneers in the 21st century should ensure that they consider and attempt to articulate possible unintended consequences of innovations they introduce and develop in or for the Metaverse.

Considering the unintended consequences of potentially well-intentioned action is essentially an ethical framework first articulated in the context of science and technology in a 19th-century novel written by a 14-year-old girl. Two centuries before Swedish teen Greta Thunberg addressed the 2018 United Nations Climate Change Conference (Wikipedia, 25 January 2022), a teenage Mary Shelley wrote her allegorical masterpiece *Frankenstein; or, the Modern Prometheus.* In this novel, published in 1818, Shelly introduced the idea that scientists should anticipate and take responsibility for the social consequences, including the unintended impact, of their discoveries, however well intentioned.

The Metaverse in a Post-Pandemic World

To articulate and assess the likely consequences of the Metaverse, it is helpful to consider the Metaverse in a post-pandemic world in a two-part temporal fashion: the near-term (next decade) and the far-term (beyond the next decade). In the near-term, the Metaverse will increasingly feature compelling high-resolution 3D, 360 visual, and aural immersive and interactive environments accessed primarily via head-mounted display (HMD) or AR/MR eyewear, or in 2D form via handheld, mobile devices, or laptop computers. Platform-independent accessibility will be a fundamental design feature so that disability or technology preference will not be limitations in the Metaverse (Brodsky, 4 February 2022). Secure blockchain-based cryptocurrency will play an increasing role as the financial transactional basis of the near-term Metaverse. The rollout of 5G will support increasingly wide-scale, mobile, and frequent public engagement with the Metaverse, especially in urban settings, taking access into the outdoor physical world beyond typical Wi-Fi-enabled venues, or even on coming Wi-Fi 6 or Wi-Fi 7 platforms (White, 2022; News18, 2022). Geo-tagged, location-based Metaverse experiences via AR and MR will be important complements to indoor VR-based Metaverse experiences. Commerce will likely thrive in the near-term Metaverse, as corporations follow the "if you build it, they will come" *field of dreams* model of innovation. Internet service providers such as Verizon might tempt users to enter the Metaverse by offering packages such as "free" or advertising-supported broadband Internet service if the user agrees to enter via the company's Metaverse portal. Such incentives have proven effective in the era of streaming media where advertising-supported video streaming (i.e., advertising video on demand, such as PlutoTV) has proven successful and growing rapidly alongside subscription video on demand, such as Paramount+).

In the far time, the Metaverse will increasingly feature advanced visualizations, spatial sound, haptics, and perhaps even more sensory engagement and envelopment that will be almost indistinguishable from live action video and even physical reality. One might think about the differences between early CGI movie animation in the early 1970s and the advanced CGI animation produced a half century later by DreamWorks and others. Videogame producers are on a similar path. The first 3D-rendered movie, or CGI animation, produced and incorporated into a 1972 movie featured a simple rendition of its creator's hand, that of Ed Catmull (Ultimate History of CGI, 2018). Catmull later founded Pixar.

In contrast is a CGI animation produced for the Disney movie *Encanto* in 2021. Among other things, the texture of the skin of the virtual characters is visually almost indistinguishable from high-resolution videos of human skin (Chryststomou, 2021). VR and other immersive content in the far-term Metaverse may feature animations that are interactive and appear as authentically reality based as any live action video shot in 8K in 2022, as featured in the 2022 Beijing Olympics broadcasts and VR experiences (Fried, 2022). The 2022 introduction of generative AI may further fuel this.

The far-term Metaverse may be accessible without the encumbrance of an HMD. Stephenson calls this the "open Metaverse" and views it as likely (Roberts, 2023). As such, the far-term Metaverse may offer UX that so closely parallel the real, physical world that virtual classrooms, workplaces, and news and entertainment venues will be comfortable for virtually anyone.

There are several potentially significant advantages, especially for news, if users can access an immersive form of the Metaverse without having to don a headset. These advantages include the fact that many users do not have a headset, and some simply would prefer to not wear a headset. And, donning a headset may limit users' mobility and therefore limit the potential of a mobile Metaverse experience. Not having to don a headset could also help make the Metaverse more widely available and mitigate the problem of the digital divide. Far-term Metaverse platforms will feature full user body tracking and display, giving the user the ability to see and feel not only their hands but also their entire body, and interact in a more physically realistic manner via all their limbs.

One possible option for an HMD-free Metaverse is the high-resolution, 8K 3D stereoscopic display (Crewe, 2022). Already in prototype form, such a display can present the Metaverse in compelling form on the desktop, and eventually on handheld devices. High-resolution devices are also shown to reduce or eliminate user discomfort or eye strain (Mellon, 2023). Another likely far-term Metaverse headset-free interface is the volumetric display. Volumetric displays project 3D, 360 visualizations, including video, into a 3D space. Typically, this is in a clear container placed on a table or similar arrangement, but larger-scale, room-based volumetric displays are possible. Volumetric displays often are designed using holography. Some notable examples have been used in museums or other exhibits, or concerts and other venues, but they are expensive and interactivity can be limited. Volumetric tensor holographic displays provide not only 3D interactive experiences but haptic, as well. Tensor holography is generated by computer rather than by light, and consequently can generate 3D imagery in virtual real time and thus present motion rather than the static images of light-based holography. Tensor holography could be displayed by a smartphone (Ackerman, 2021).

Cost is likely to fall over time, given Moore's Law (1965), and volumetric, holographic displays for the home or office may become more viable. Holographic displays that include interactivity are in development. PORTL M is one such holographic volumetric system. It acts as a two-way hologram communications device, displaying interactive 3D, 360 visualizations. To what extent the content will feature synchronized spatial sound and interactive haptics is unclear. And at a cost of $2,000, PORTL M is far from mainstream public adoption, limiting the potential use in Metaverse journalistic content (Brodsky, 28 January 2022). As of 2023, most HMD-free systems are still in early development, or are limited to arcade or other site installations, and are not typically available to home or office users. But they do point a way toward a potential far-term Metaverse for the masses and thereby news use.

Considering the Consequences

Among the most-often articulated corporate visions for the uses and functions of the Metaverse, especially in the near term, is an immersive virtual world to meet friends, family, and colleagues. Likewise, this corporate vision is for the Metaverse as a place to work and collaborate with others, to enjoy virtual life, and to create and partake in virtual experiences. Meta highlights this in its promotion of *Horizon Worlds*, emphasizing it as a place for "creators." Meta reports that as of February 2022, users created 10,000 worlds and more than 20,000 users joined a private Facebook group for creators (Tracy, 2022). This could help lead to the development of what is called the creator economy or an economic model built on revenues derived from users who create immersive content in the Metaverse. These could be freelance journalists.

Gartner predicts that by 2026 a quarter of the public will spend at least an hour a day in the Metaverse (Gartner, 2022). As such, the emerging Metaverse is likely to become a place where corporations can generate billions, maybe trillions of dollars, although how much of this is hype is hard to say. JPMorgan Chase predicts the Metaverse will impact every sector of the economy and generate extraordinary market value (Lenihan, 2022; JPMorgan Chase, 2022). PwC projects that VR and AR may add US$1.5 trillion to the global economy by 2030 (Hernandez, Likens, Priest, Korizis, Panjwani & Rivet, 2022). PwC projects that US$294.2 billion will be for immersive training or professional education. PwC states there are four major reasons for this, several of which speak to the major user conditions noted above. First, setting up and using VR and AR user platforms is significantly easier than just a few years ago. In 2016, when Oculus VR was released, the Rift VR system was complicated, space-intensive, and costly (at least $1500). The setup was complex and few users could install the system without professional technical assistance. By 2022, the setup of the Quest, the Rift's descendant, had become relatively simple, much more like setting up a new smart television set or a Roku set-top TV (over-the-top, OTT) box. AR systems are extremely easy to set up and use, with smartphones and tablets being the main platforms. Users only need to install a downloadable app. Second, VR and related XR systems are much less costly than systems in 2016 (a decrease of about 80%). By 2023, consumer-friendly VR systems typically cost about $300, although there are still expensive high-end devices. Third, VR platforms are untethered. Systems in 2016 were connected by wire to a dedicated VR-enabled computer, limiting user mobility and comfort. Contemporary VR systems are wireless, or untethered, and not only facilitate user ease and comfort but support better system security and remote management. Fourth, the requirements of contemporary workspaces have evolved, partly in response to the COVID-19 pandemic and the need to support worker social distancing and remote work. Classroom training is complex and raises concerns over health not relevant to workers at a distance. VR system-based training mitigates some of these concerns. These four reasons are pertinent to journalism, both for possible training of reporters to work inside the Metaverse and for potential users of Metaverse journalism.

It is likely that journalism and media will be present in a substantial way and in widely varied forms in the early Metaverse. Some news enterprises and media organizations already have begun to stake a claim in the Metaverse. For instance, the BBC, Gannett, and *The New York Times* are among the news organizations already invested in creating immersive, VR-based content experiences (Doyle & Gelman, 2016). This has led media observers such as Thomas Seymat (2021) to propose that the Metaverse might even be the engine that "saves" journalism, an industry that has been in disruption, even economic free-fall, in recent decades fueled by the rise of the pre-Metaverse Internet. Metaverse News is a website devoted to news about the Metaverse (2022). It reports on everything from the concept of the Metaverse to a wide range of possible Metaverse components and concerns, such as nonfungible tokens (NFTs), games, and legal issues. With artists such as Ariana Grande under contract, Universal Music has inked a deal to generate revenue selling NFTs of its artists inside the Metaverse (Reuters, 17 February 2022).

The mediated experiences users will engage in the Metaverse are likely to vary widely, from immersive news to cinematic VR where avatars or digital twins observe and enter as participants in virtual narratives and act as cyber reporters and virtual eyewitnesses to news events. Cryptocurrency may become a dominant economic engine, fueled by the advance of the Metaverse. AI agents will act as user guides. Moreover, AI will play a key role in tracking users within and without the Metaverse and will generate real-time personalized Metaverse experiences. Social VR platform VRChat has introduced face tracking for avatars (Melnick, 2022). Billions of avatars with human real-life twins will be complemented by potentially even more bot-based avatars, and humans may be unable to tell the difference.

There will be many alternative Metaverse platforms and they will vary in some important ways. Most will offer a free version funded by in-Metaverse purchases as well as AI-driven personalized advertising, marketing, and news. But there also will be premium versions with much higher quality immersive and interactive content, including news, that is largely free of advertising content.

The potential benefits of the near-term Metaverse are numerous. Among the most intriguing is the potential to enable persons to transcend the physical or temporal boundaries of their current situation, however defined, and experience the wider world, or even to go beyond the known world into an imagined virtual realm (Kahn, 2022). The Metaverse could do this in many ways, such as by enabling virtual travel and cultural experiences, attending concerts and shows, participating in virtual games, including news games, or advancing the streaming entertainment industry into more immersive forms, and facilitating immersive retail transactions (Marson, 2018).

Benefits for journalism, media, and society include the generation of immersive experiences that through high levels of user presence can lead to greater empathy and understanding. This might help reduce the social, cultural, and political division that increasingly characterizes many democratic societies. One possibility for news media is to design simulations within the Metaverse that enable users to learn, enjoy, and otherwise experience the real world or models of the real world in a highly convincing, persuasive, and informative fashion, and by placing the user into a first-person perspective that is not necessarily their own. Such simulations could enable users to fully understand how a pandemic occurs, how a virus spreads, how masks work, how vaccines work and why they are so important, and how all these factors intersect with identity. Climate change simulations could model the world in a highly personalized fashion to allow each user to understand how climate change affects them in their local community, and how reducing the human carbon footprint is essential. Modeling the climate or energy consumption of the Metaverse itself also is possible.

The consequences of extensive personalized advertising and marketing featuring micro-targeting via AI and user data tracking will deeply impact the Metaverse. Surveillance capitalism in the extreme will likely develop. Privacy considerations, especially regarding children and youthful users, will be paramount. How Metaverse journalism responds and proactively protects user privacy is a critical ethical consideration.

Computer scientist, mathematician, and artist Jaron Lanier coined the term virtual reality as his company, VPL Research, developed the first consumer VR headset in 1987 (Internet Society, 2017; Virtual Reality Society, 2018). Now a Microsoft research scientist, Lanier (2013) has since written that his vision for VR is in dramatic contrast to what it is becoming in the corporate-dominated Metaverse (Microsoft, 2022). He imagined a virtual realm largely free of commercial influence. Lanier's vision of VR was an almost magical place where creativity and content could flourish in an interactive, immersive, and empowering realm. Lanier described his early haptic experience in VR, before its increasing corporate domination. He used a data glove to virtually touch and interact with a huge ship at sea in the harbor near Seattle, WA. This is an apt metaphor for journalism and society with regard to the Metaverse in 2023. If the Metaverse is imagined as a giant tanker or container ship, a minor "touch" or nudge (Sunstein, 2015) today could shape its trajectory tomorrow in vital ways that place ethics and social benefit in the forefront and ensure user data exploitation is highly constrained and regulated. Waiting to attempt that contact with the ship Metaverse would make changing its course in a significant way nearly impossible or require much more force, energy, or resources. Journalism organizations and journalists can play a key role in creating the "nudge" that could shape the long-range path of the Metaverse. By reporting thoroughly about the Metaverse and within the Metaverse, quality journalism can help to shape public understanding of the full range of problems and opportunities the Metaverse presents.

Conclusions

The Metaverse is likely to bring with it important implications for journalism, media, and society, perhaps more disruptive than even the Internet to date. As the Metaverse continues to evolve and develop, it is critically important that journalism and media scholars conduct independent research about those implications, including the nature and consequences of the Metaverse for the shape of news and news engagement. Establishing an agenda for research on the nature and consequences of the Metaverse, especially Metaverse journalism, is a vital next step for media scholars interested in the future of digital communications. For the news media, the challenge of the Metaverse is to develop immersive journalism content that brings society closer to the truth. The potential for journalism is to advance clarity of understanding on matters of public importance and thereby benefit the democratic process as it transforms into an immersive digital estate.

Chapter 2

PRINCIPLES OF JOURNALISM IN A VIRTUAL WORLD: ENSURING THE PURSUIT OF TRUTH

This chapter addresses the question of whether the principles of quality journalism in a virtual world are the same as in the real, or physical, world. This chapter asserts that there are key differences. Kovach and Rosenstiel (2007) argue that the first "obligation" of journalism is to report the truth. Fulfilling this obligation enables journalism to serve the public interest. This foundational principle still applies to Metaverse journalism. But several other principles take on heightened importance in the Metaverse. It is critical to outline these differences early in the book because the chapters that follow build upon this framework. This chapter articulates three core principles that are particularly vital to quality journalism in a virtual world. These are editorial independence, ethical practice, and full transparency in the pursuit of truth.

Although these principles pertain to real- or physical-world journalism, they rise to paramount importance in the Metaverse. The digital, coded, and virtual nature of the Metaverse makes this so. Moreover, the essential principles of real-world journalism, such as ensuring accuracy, fairness, and inclusion, minimizing bias, and providing context, still apply inside the Metaverse. But the three principles identified above are of uniquely high priority in the emerging virtual world of the Metaverse.

Editorial Independence

Editorial independence means those producing journalism in the Metaverse must be independent from, or not under the authority of, those who control the Metaverse, particularly with regard to story selection, reporting, and presentation. Without this independence, journalism in the Metaverse will be suspect and little more than public relations or marketing, or even propaganda, as the Metaverse is likely to be entirely privately owned and operated or owned and controlled by international state players or aligned entities. A primary reason independence will be essential in journalism in the Metaverse is to avoid any conflict of interest, particularly financial, that may arise if journalists reporting in or on the Metaverse have a business or economic stake in the Metaverse or the companies or persons who control or design it. In the Metaverse, cryptocurrency is likely to be the main engine that drives its development. Journalists and the organizations that they represent may be particularly susceptible to the influence of the economics of crypto or the fiscal underpinnings of the Metaverse. The problem of news media economics has become acute in the 21st century, as long-held business models and economics of the media have collapsed and have required reinvention. As such, the pressure of digital economics likely will be profound on journalists and news media organizations operating in or about the Metaverse. News media may be major contributors to the development and growth of the Metaverse. As is discussed in further detail later in the book, news media have been leading creators of content that populates the Metaverse, including immersive journalism and cinematic VR. These content forms are expensive to develop and news media are under enormous pressure to fund their development. Journalists need to be insulated from this pressure so that they can produce quality journalism in and about the Metaverse without any possibility of a lack of integrity or perception of it. Independence of journalism in and about the Metaverse will help to ensure that the public, especially users of the Metaverse, can trust the integrity and honesty of what the news media report.

The rise of mis- and disinformation during the COVID-19 pandemic as well as more widely since the 2016 US presidential election has highlighted the corrupting influence of a failing system of trustworthy news and information in the digital space. This problem will only grow more intense in an era of the Metaverse. To the extent that robotic systems of content generation and distribution (i.e., bots) grow in the Metaverse, as AI and especially generative AI grows, the public will need to know that they can believe the journalism they see, hear and otherwise experience inside and about the Metaverse. Research shows that these issues pertain to real-world journalism but they become more acute online and exponentially so in the immersive Internet, or Metaverse.

Editorial independence also means being free from governmental interference or influence. Moreover, the Metaverse will be a global platform that largely transcends the boundaries, geographical or political, that exist in the physical world. This means independence from any governmental entity or aligned entity that might seek to shape or control editorial decisions and content generated and published by a news organization. It becomes a cloudy area when questions about the potential regulation of generative AI that might be in the employ of news media.

Rosen (1999) is among those who have commented about the importance of independence in journalism. He has argued that journalists should be "public intellectuals" and are as such accountable to their audiences. This is a particularly salient point for journalists in the Metaverse where the status of the traditional public square and public sphere are unsettled.

Rosen has designed and launched a news service and site titled *The Correspondent* (https://thecorrespondent.com/), based on the Dutch *de Correspondent* (https://decorrespondent.nl/). Subscription-based and advertising-free, *The Correspondent* is intended as an independent news platform. It could be a model for such a news service produced for the Metaverse.

Duval has considered the role of newspapers, or the press, in shaping public opinion during the American Revolution (2016). She has articulated the critical importance of journalistic independence in an emerging democratic society. Whether the Metaverse is or should be considered a democratic space, it will exist and operate in a context in which democratic institutions and principles operate and independent journalism will be essential to the ultimate shape it takes and values it reflects.

As a scholar of media studies, Herman was a leader in the analysis of a propaganda model of the media. This model asserts that the media serves the interests of the powerful, including individuals and institutions, rather than the public. Collaborating with Noam Chomsky, Herman argued that journalistic independence is essential to counter this systemic bias. The Chomsky and Herman propaganda model is particularly important and influential in shaping the understanding of media and propaganda in this regard (1988). In countering propaganda, journalism promotes a more democratic media system. In the Metaverse, the propaganda model is particularly likely to play a significant role and it will be essential that independent journalism act as a check against it. Lasswell (1933) was among the first to articulate the role that media can play in propaganda and subsequent work builds upon and extends this foundation.

As an author of numerous books on media and journalism, McChesney, in collaboration with Herman, has argued that corporate ownership and media consolidation pose dangerous threats to journalistic independence (Herman & McChesney, 2001). The threat lies in the fact that commercially owned and profit-motivated media, including journalism, is motivated to advance capitalism. In presenting this argument, McChesney makes a case for greater protections, including legal and structural, for an independent press. In the Metaverse, how to protect journalistic independence may require both legal and structural mechanisms. If journalistic enterprises inside the Metaverse are similarly motivated to support capitalism, then their ability to report critically on the nature and function of the Metaverse will be compromised.

Ethical Practice

Ethical practice must be the foundation for journalism in and about the Metaverse. Although the Metaverse may be regulated and subject to legal frameworks from outside its technological parameters, its boundaries and characteristics will likely shift much more rapidly than external regulators and lawmakers will be able to address in a timely manner. The moral compass of independent journalism will be a critical mechanism to make sense in near real-time of the shifting digital sands of the Metaverse. Ethical practice will be at a premium for journalism in the Metaverse. This heightened need for strong ethical practice rests on at least three factors that will prove pivotal to successful Metaverse journalism. First, the Metaverse has a high potential to be corrupted by economics as corporations with dubious ethics track records take a lead role in developing and dominating the Metaverse. Second, user data will likely play a key role in the functioning and financing of the Metaverse. User data have been at the economic foundation of digital platforms and it seems unlikely this will change, unless it accelerates even further in the Metaverse. Ethical practice also means ensuring full accessibility for all persons regardless of disability. For journalism, this means featuring a set of design and user interface practices that maximize access across modalities. Protecting user privacy will be paramount in any sustainable journalistic endeavor and especially so when a full spectrum of users are accessing and engaging with that journalism. Third, barriers to entry in creating journalism in the Metaverse will be relatively low. And any news organization (or journalist) that fails ethically will likely quickly lose user trust and disintegrate.

In reporting in and on the Metaverse, journalists and news organizations should adhere to at least three core ethical principles. First, they should consider the consequences of what they report. Second, they should take steps to ensure the protection of user privacy. Third, they should endeavor to respect the intellectual property (IP) rights of those in or represented in the Metaverse.

Ethical practice in Metaverse journalism takes a different, more fluid shape than it has had in the real world, at least up until recently when the pace of technological change has been so fast and far-reaching that the rules that govern ethical journalistic practice have become increasingly obsolete.

The shape-shifting nature of the Metaverse makes it particularly important that journalists and news media consider the consequences of their journalistic practices and base their news decisions at least in part on a consideration of those consequences. A classroom visit by a special guest more than two decades ago helps illustrate the ethical imperative of considering the consequences.

Considering Consequences in Decision-Making

Mr. Pete Seeger was a great singer, songwriter, and civil rights and environmental activist whose life had a transformative impact on the world. In the 1990s, I had the good fortune to become his friend. One day he accepted my invitation to visit my journalism class at Columbia University to talk about storytelling. I offered to arrange a car service to bring him from his home to my class, as it was a journey of more than 50 miles and he was in his 70s. But he declined. He understood the harmful consequences of burning fossil fuels and he insisted on taking public transportation to minimize his carbon footprint. The class was in the evening, and I was sitting in my office getting ready for the evening's seminar. When I looked up to see Mr. Seeger walking through my office door I felt relief. He was dressed as usual and utterly at ease in his blue jeans, flannel shirt, and baseball cap, with his banjo case over his shoulder. After saying hello, we walked together to class. The topic that evening was journalism, storytelling, and the implications of new technology for reporting. It was a topic Mr. Seeger had deeply considered and we had often discussed. He'd even written a foreword about it for a book I wrote about media in the digital age.

I suspected my students might not know exactly who Pete Seeger was, so I had prepared them in advance. They learned about his early life interest in becoming a journalist, that he'd written multiple books. They knew he was inducted into the Rock 'n Roll Hall of Fame for songs he'd written and performed, including masterpieces such as *If I Had Hammer*, a song of freedom, love, and justice, *Where Have All the Flowers Gone*, an anti-war song of peace, and *We Shall Overcome*, the civil rights anthem. They knew Mr. Seeger was a champion of the environmental movement, helping protect the Hudson River since at least the 1960s. They knew he was nearly killed by rioters opposed to his performing with Mr. Paul Robeson, the legendary actor and civil rights champion, at a concert in 1949 in Peekskill, NY, in support of labor rights. The students were engrossed in Mr. Seeger's blending of an in-class musical performance with his explanation of the importance of storytelling. They admired his banjo, inscribed with the message, "This machine surrounds hate and forces it to surrender." What they didn't expect was his familiarity with many emerging technologies and their potential implications for storytelling in journalism. Mr. Seeger told the students that they as journalists could use new technology to transform news gathering and storytelling. They could help to transform journalism in a time of disruption. But, Mr. Seeger cautioned, they should first ask one simple question: whether they should. Ethics are paramount, Mr. Seeger asserted. It is essential to consider the consequences of any action, inaction, or decision. Journalists should consider the benefits and the harms, and for whom, Mr. Seeger added. We should reflect on how the use of these new technologies in journalism will affect understanding, personal privacy, diversity, inclusion, and equality.

Mr. Seeger and I recalled to the students that our classroom was only a short walk from the laboratory where Dr. Enrico Fermi worked to invent the atomic bomb. Fermi's research led to the Manhattan Project and the bomb dropped on Hiroshima, Japan, at the end of World War II. The core ethical question is whether Fermi should have helped to develop nuclear weapons. Reflecting on the Fermi case, and whether to use new technology in journalism, my students thought about the ethical conundrum. By protecting user privacy, ensuring accuracy, and prioritizing the pursuit of truth, the students made ethics their guide in the use of new reporting technologies. In this ethical way, the students employed 360-degree video to report on the efforts of the Irish Lesbian Gay Organization to March in New York City's annual St Patrick's Day parade, from which they'd been banned and subject to arrest. The students used this technology to provide an immersive look at the story of Amadou Diallo, whom police killed late one night as he stepped innocent of any crime from the vestibule of his apartment building in the Bronx. My students used AR to create a situated documentary about Fermi's research in New York. The Fermi narrative included an interactive and simulated view of the devastating consequences if terrorists were to detonate a nuclear explosion at the Empire State Building. Mr. Seeger's simple question should guide each journalistic action in the Metaverse.

The Metaverse is likely to be largely corporate owned and there will likely be little to no public space. Consequently, any user will be at a substantial power disadvantage. Power imbalance is a key principle in considering the consequences of a story and how to report it (e.g., protecting the identity of a sexual assault survivor). Power differential in the Metaverse makes it imperative that news media follow a consequences consideration first model for all stories. The question of whether the public is likely to be harmed by the reporting of a story, or whether there is likely to be a benefit should guide whether and how to report a story in the Metaverse. Prioritizing a consideration of the consequences of publication will help to ensure the sustainability of journalism in the Metaverse.

Anticipating the Unintended Consequences

Inequalities fueled by platform capitalism likely will grow exponentially in the Metaverse. Multi-million-dollar virtual property sales as well as NFT transactions, potentially via AR, VR, and other forms of XR, have taken place in the Metaverse. In November 2021, a piece of virtual property sold for $4.3 million in Sandbox, a popular Metaverse platform (Mozée, 2021). How news media organizations enter and establish a presence inside the Metaverse may be critical to ensuring inequality does not thoroughly undermine the potential good of or benefit derived from Metaverse journalism.

In parallel to economic inequalities, the digital divide is likely to surge in the Metaverse. Those with the means will be able to enter, share, and reap the economic and other rewards of a high-end Metaverse. Those who lack the means either will be denied access or relegated to an inferior version where problems of misinformation abound and surveillance capitalism runs rampant. In fact, supercharged mis- and disinformation are likely, particularly in an age of powerful generative AI platforms. The Metaverse could destroy the last shred of user privacy. Metaverse journalism needs an open and secure pathway for all where privacy and other user rights are protected.

Addiction and other mental health effects of the Metaverse are also likely. Yet, research shows that VR, a key element of the Metaverse, can be very effective in treating mental health problems, including post-traumatic stress syndrome. But such benefits are likely only if the technology platforms that build the Metaverse design it ethically, putting people before profit (Smith, 2022). In other words, designing and operating the Metaverse to feature the potential mental health benefits of VR may be expensive and require big tech to sacrifice revenues. It is unclear if they would likely commit to such a path, given many of their past business practices. How news media confront these challenges will be a test of their ethical mettle in the Metaverse.

Identity issues similarly may emerge as significant problems of the Metaverse.

Though virtual, the Metaverse may be a dangerous place. Criminal activity may be a major problem in the Metaverse. In fact, research shows that sexual assault in virtual worlds occurs and is a potentially major problem for the development of the metaverse (Mergerson, 2018; Cho, 2022; Davies, 2021; White, 2022;). *Horizon Worlds* has proven a draw for child users and predators may be lurking (Heath, 2022; Oremus, 2022). Meta allows children as young as 10 to use the Quest, with parental permission. Human trafficking is also reported to be emerging as a potential problem in the Metaverse, and as such will be another important arena for investigative journalism (Edwards, 2022). Just as X (formerly Twitter) hen to be a haven for child abuse and sexually exploitative content, the Metaverse could become even worse unless strong guardrails are in place (Belanger, 2023). How to prevent such content while protecting freedom of speech and press is a vexing challenge. Physical risks to user safety also are present in the Metaverse. Even accidental injury is a major concern in the Metaverse; insurance claims due to injury incurred while in VR grew by nearly a third (31%) in 2021 (Smith, 2022). Advanced AIs may be a corrupting force in the Metaverse, just as bots have been for social media users. Bots and other AI-powered agents may sew discord, disruption, and manipulation and do so in a manner that appears completely real. Deepfake news video programming is increasingly commonplace in 2023 and increasingly hard to detect. "In one video, a news anchor with perfectly combed dark hair and a stubbly beard outlined what he saw as the United States' shameful lack of action against gun violence," report Satariano and Mozur (2023). "In another video, a female news anchor heralded China's role in geopolitical relations at an international summit meeting," Satariano and Mozur state (2023). The on-screen speakers were anchors for a news organization called Wolf News. But the anchors' voices were slightly out of sync with the movement of their mouths, Satariano and Mozur add (2023). Their faces were somewhat pixelated. Closed captioning contained grammatical errors. In reality, the anchors are computer-generated avatars produced via AI. Pro-China bot accounts distributed videos of these news anchors in 2022 on Facebook and X. It is the first case identified in which deepfake news videos have been distributed as part of a state-affiliated disinformation campaign. News media must ensure their coverage of the Metaverse does not minimize these issues and in fact highlights the potential occurrence of immersive deepfake news content in the Metaverse.

Largely corporate dominated, the many varied Metaverse platforms will increasingly be owned by and profitable to a handful of mega corporations such as Meta, Nintendo, and Apple. In such a future commercial multi-platform Metaverse, there may be little place for common space (Aristotle, 330 BCE; Neacsu, 2022). The public village square may vanish in a corporate-owned, controlled, and expensive Metaverse. It is not clear whether it is possible for a nonprofit Metaverse to be designed, built, or sustained. Quality Metaverse journalism may not be able to exist without such a virtual public square.

Protecting Privacy

Privacy soon may be little more than a distant memory regardless of the Metaverse (Steele, 2022). Although Apple CEO Tim Cook has publicly affirmed his company's commitment to protect user privacy online or off, unless the Metaverse is highly regulated, consenting to continuous data tracking and corporate exploitation may be a virtual requirement of admission to the Metaverse, particularly for those seeking free admission (Steen, 2022). This may be subject to regulatory frameworks such as the General Data Protection Regulation (GDPR) in Europe which has considerably stricter privacy and data collection and disclosure requirements (e.g., per cookies) than in most of the United States, outside of California. Journalists will need to strike an ethical balance between respecting user privacy boundaries and the public's right to know about the flaws in the Metaverse.

Privacy protection is a foundational component of ethical journalism in the Metaverse. Privacy means the capacity of an individual or group to secrete themselves or information about themselves, including their data. The right to privacy includes giving an individual (including an Internet or Metaverse user) the ability to present themselves selectively. In the physical world, people employ a variety of techniques to present themselves selectively including via clothing, makeup, and the like. In the digital realm, there is similarly a variety of tools available to users to control or limit what they present about themselves or how. In the Metaverse, this selectivity is especially evident in the user's avatar. But there are other tools available as well. The right to privacy implies the right to be free from public scrutiny or undesired attention or interaction. In the Metaverse, one form this takes is a virtual protective shield users can invoke to restrict any attention or virtual contact. But this is privacy protection from other users, not from the platform itself. Confidentiality, or the ability to secure information from unauthorized access or disclosure, is often closely linked to privacy. Privacy is similarly closely tied to the notion of anonymity. Anonymity means not being identified or recognized. Ohm is among the legal scholars who have examined the intersection of technology, privacy, and anonymity. Digital platforms have used sophisticated techniques to "transform sensitive into non-sensitive information through anonymization—the elimination of personal identifiers like names and social security numbers," Ohm explains (2009). This approach can satisfy regulatory requirements. "Google anonymizes data in its search query database after nine months and health researchers aggregate statistics before publishing them," Ohm adds. But developments have revealed that anonymization has potentially very limited power to protect privacy. "America Online and Netflix each released millions of anonymized records containing the secrets of hundreds of thousands of users. In both cases, to the surprise of all, researchers were able to 'deanonymize' or 'reidentify' some of the people in the data with ease" (Ohm, 2009). Privacy therefore is a complex and multifaceted concept. It brings with it significant social, ethical, and legal considerations. Privacy is among the most fundamental of human rights. It is protected by law and regulations across the globe.

Journalism in the Metaverse must question the use of anonymization and should critically investigate digital platforms that use this technique under the premise that it protects user privacy.

Among the scholars who have studied the intersection of technology, law, and policy is Kevin D. Werbach. His work has particularly focused on privacy, security, and innovation, areas that increasingly converge in the Metaverse. Werbach's examination of blockchain (2018) and how it may serve as an architecture for trust is especially salient. Sullivan (2022) describes the nature of the blockchain as, "a public ledger system that exists only on the Internet. It uses a complex cryptography system to ensure that everything written to it ('blocks') is verifiable." Web 3 is the term often used to describe Internet service that uses decentralized blockchains, and thereby can support WebXR content, including journalism (Roose, 2022). It will be at the core of the Metaverse. Blockchain likely will serve as an underlying foundation for secure transactions in the Metaverse and can protect the security of digital news content and user data and thereby privacy. For journalism, the convergence of blockchain, security, and privacy represents an area of importance for reporting on the Metaverse as well as how journalism develops within it.

In the Metaverse, privacy in all its dimensions will be a key consideration. Users' capacity to seclude themselves, speak anonymously, conduct confidential communications, and control their presentation all will be vital but contested, and bring substantial challenges to journalism operating within the Metaverse. Criminal activity such as fraud inside the Metaverse is a good case in point. In an age of brain-tracking wearable technology, how to protect user privacy while not enabling fraud or other criminal activity to proceed under a protective shield will challenge regulators, platforms, and journalists seeking to investigate it.

Nissenbaum (2010), a professor at Cornell Tech, has written extensively on privacy and values in information technology. She makes a strong case that new digital technology has generated major threats to privacy, as it has enabled user tracking and more. She argues that protections should be in place for appropriate flows of user information, as all persons will exist in some fashion within an evolving information environment. This is especially the case with regard to the Metaverse, where flows of information will be a largely defining quality of the user experience.

Daniel Solove is Eugene L. and Barbara A. Bernard Professor of Intellectual Property and Technology Law at Washington University Law School. Through his research, he makes a compelling case for the importance of privacy law and data protection. In his book *Breached* (2022), Solove presents evidence that current laws and policies generally in fact contribute to data insecurity.

This may become a particularly profound consideration in the Metaverse. For journalism, it is not only a topic for reporting and investigation but also an important standard for how news media engage users. Moreover, failure to protect user privacy is increasingly likely to result in regulatory scrutiny (Selinger, Polonetsky & Tene, 2018). Even substantial fines may result, such as was the case in early 2023 when the Irish Data Protection Commission levied a 390 million euro fine against Meta's social networks Facebook and Instagram (Bracy, 2023).

Respecting Intellectual Property Rights

IP rights are important in all forms of journalism but they take on particularly pronounced and acute importance in the Metaverse. This is a consequence of the fluid nature of the Metaverse and the manner in which it is driven by both data and algorithm or AI, and also the fact the public, or users, will be essentially partners in the creation and curation of journalism in the Metaverse. As such, the full delineation of IP rights will be key. For journalism, this will mean using blockchain to secure those IP rights and be able to provide a certificate of authenticity as well as the provenance of what is being reported, by whom (possibly compromising privacy), and who is responsible for a news product, such as a story, interactive graphic, or immersive video, including for every fact, source or other component of a news item. This derives from the concept of hypertext Ted Nelson introduced in 1965 (Nelson, 1965). Among IP issues with regard to the Metaverse is the infringement of trademarks, patents, copyrights, and plagiarism, either in the form of a human copying from a bot or vice versa (James, 2023). Scholars conducting research in this arena include Tilleke and Gibbins (2022), Bahuguna (2022), and Deshmukh (2022). Bregman, Gettleman, Crisman, and Hemmer (2022) have examined antitrust and international rights issues in the Metaverse. In one case, an innovative college student has designed an app that can detect if text was created by AI (Bowman, 2023).

Full Transparency

Transparency in journalism means making the news process visible and accessible to the public. Making the provenance of a news item available to the public is an essential element of this process of providing full transparency. Transparency in the Metaverse will particularly rest on a foundation of data. Because the Metaverse is a digital world, data (and datafication) will be at its foundation, and as such, data journalism will be essential to news production and content in and about the Metaverse. Transparency in the manner in which those data are collected, utilized, reported, and otherwise manipulated will be key. Other types of reporting will be important as well, but through data analysis, this reporting can be put in context to make it substantially more meaningful and not simply anecdotal. Data collection will be continuous and widespread if not ubiquitous. Data will take many forms, from transactional (e.g., NFT sales) to user tracking. User tracking itself will be a multidimensional matter, including monitoring user activities such as attending virtual events and micro-movements, such as facial gestures. As such, strict rules will need to be developed that ensure journalistic excellence and also protect users from unwanted scrutiny by either the platforms or news media. Journalism must implement and follow a strict policy of data responsibility in both reporting and user engagement.

News media must deliver high transparency in news content and practices or face gradual diminishment. Providing transparency in news content and practices is essential to build public trust. As a substantial portion of the population of users of the Metaverse, gamers provide a case in point. Research shows that gamers have become disgruntled and lost trust upon learning they were playing online games against bots rather than vying against what they thought were human opponents (Needleman, 2022). Although transparency is important in all journalism, it is especially salient in the Metaverse. The most important reason for this is that the Metaverse exists only as code. Without heightened transparency, the public will suspect that all Metaverse journalism is compromised.

Scholars studying transparency in journalism have examined its implications for trust, credibility, and legitimacy, all of which are vital issues for journalism and the Metaverse (Haapanen, 2022). Scholars studying normative considerations of transparency have advanced the value of openness in journalism (Koliska, 2015). Designers of human–computer interfaces have tested the potential of online news platforms to increase trustworthiness through the use of transparency cues (Bhuiyan, Whitley, Horning, Lee & Mitra, 2021). These cues include the presence of source materials, avoiding anonymous sourcing, verification of content, and posting corrections to erroneous reporting. With the advance of AI, it will be essential to clearly label content generated by AI in part or in whole. The integration of transparency cues in Metaverse journalism can facilitate the development of user trust.

Conclusions

The combination of these core principles, editorial independence, ethical practice, and full transparency, will fuel journalism in its pursuit of truth in the Metaverse. Their combination can help mitigate potential bias in reporting within the Metaverse and build user trust. The pursuit of truth has never been more contested or more important. The layered nature of truth, beginning with data, facts, and context, will be an essential part of the Metaverse, a digital world of code, and how it is understood and reported. Journalism will be essential in providing a reliable, truthful source of the interpretation of that code and its manifestation for the Metaverse's users, observers, and other interested parties. Bias in news has been a vexing problem journalism in the real world has faced and often failed to prevent. It has often undermined efforts to present the news objectively, or without favoring one viewpoint or another, either explicitly or implicitly. Word choice, exclusion of voices, and limited perspectives are among the forms of bias that undermine the credibility of news. Scholars such as Jay Rosen (1993), Kathleen Hall Jamieson (2012), Thomas E. Patterson (2019), and Shanto Iyengar (2010) are among those who have pointed to the fundamental problem of media bias and its corrosive effect on democracy. Media shape public perceptions. Metaverse journalism must be committed to presenting news content that does not favor any group or perspective. If not, the potential for the Metaverse to develop as a democratic platform will fail.

Chapter 3

STRUCTURAL AND SYSTEMIC CONSIDERATIONS IN THE METAVERSE: REGULATIONS, ECONOMICS, AND NEWSROOM ORGANIZATION

As important as structural and systematic considerations may be in the real world, they are at least as fundamental to the Metaverse (Flowers, 1995). When Robert Kahn and Vinton Cerf conceived the Internet, its essence was the underlying digital structure. The only requirement for any computer network to connect to the Internet was that it conform to that basic structure. The digital structure likewise will shape and limit anything that happens within the Metaverse. As such, Metaverse journalism and how it functions will be subject to these same structural parameters.

This chapter outlines and critically examines the structural and systemic parameters that will define and shape journalism in the Metaverse. In particular, the chapter examines the legal and regulatory framework that constrains or enables journalism in the virtual world (e.g., how principles of freedom of speech and press may apply in the Metaverse) as well as the organizational mechanisms designed into Metaverse platforms that shape their affordances. The chapter considers the structural and systemic factors that will shape privacy in the Metaverse and how that will affect journalistic practice. The chapter examines the economic forces (e.g., revenue sources, ownership structures, system of financial transactions or currency ... likely cryptocurrency) that will shape the Metaverse. This includes examining the ownership and financial models that will shape news media entities operating within a Metaverse platform. Finally, the chapter explores the likely organizational parameters of news media within the Metaverse. This means considering the need for a news "room" in the Metaverse, or whether some other collaborative or competitive structures make more sense.

Freedom of Speech and Virtual Press

One of the significant challenges of ensuring robust journalism inside the Metaverse is balancing freedom of speech and virtual press with legal and regulatory protections to ensure an environment that supports intellectual property (IP) rights, protections for privacy, and minimizes harm from mis- and disinformation as generated by AI or otherwise. Regulating something that does not yet exist, or at least is fully formed such as the Metaverse, poses particular challenges (Robertson, 2023). Yet, forming a regulatory structure before the Metaverse is fully formed may help to shape its contours before it is increasingly difficult and disruptive to change.

Parody and satire are among the most important forms of journalism in providing criticism of public figures, particularly those with political power. Throughout history, the most effective forms of independent journalistic criticism has been in the form of political cartoons and other types of satire and parody, such as late-night video comedy that lampoons those in power. Those in power often object strenuously to such satire, as it can subject them to public illumination of harsh truths they would rather stay untold and in the dark. Public figures often object to the disinfecting power of bright illumination, whether from real or artificial sunlight. Ron DeSantis, Florida Governor (R), is a political figure seeking to roll back press freedom, and if he succeeds, the robustness of critical, investigative journalism will suffer greatly both in the real world and in the virtual world (Bensinger, 2023). DeSantis is hoping the Court will overturn the landmark case *New York Times* v. Sullivan. It established that public figures such as a governor or a CEO could not successfully sue a news organization or journalist for defamation, or libel, for an inadvertent error in reporting. Rather, they need to prove the reporter knew the reporting was false and acted with actual malice or reckless disregard for the truth. The actual malice standard is much higher than for private citizens who might be the subject of a news story. For the Metaverse, maintaining the higher standard means that robust journalism including satire can continue in reporting about public figures such as elected officials, corporate leaders, or owners of Metaverse platforms. Without this protection, vigorous journalism will be chilled to a little more than the reporting of news releases and pseudo-events (Boorstin, 1962).

X under Elon Musk offers a cautionary tale. Control and restrictions of free speech in the Metaverse may be severely curtailed not just by governmental regulators and lawmakers but by corporate owners and actors seeking to maximize profit. In 2022, Musk purchased X after declaring he would make the platform a free speech zone and that he himself was a free speech "absolutist" (Zahn, 2022). But one of his first actions was to restrict freedom of speech (Kolodny, 2022). He suspended the accounts of several comedians who had created parody accounts as well as the account of a college student whose X feed tracked the movements of Musk's private jet (Capoot, 2022).

XR developer Louis Rosenberg, who is credited with developing the first interactive AR system, argues that the Metaverse needs to be regulated and from the start (Witt, 2022). The problems seen in social media will be compounded in the Metaverse. Problems of misinformation, deepfake manipulation, and monetization that rely on exploiting user data and privacy will be in the extreme in the Metaverse and warrant the need for regulation, legal intervention, and platform governance (Gillespie, 2010). Offering some potential support for freedom of digital speech, in 2023, the US Supreme Court ruled in support of Section 230 of the Telecommunications Act that protects digital platforms such as Google from legal action arising from user content that might be libelous, dangerous, or otherwise problematic. Ending Section 230 protections would cause a potentially seismic shift in digital platform economics and content on the Metaverse. Regardless of the Court's recent decision, Section 230 may not apply to ChatBots (Cristiano & DiMolfetta, 2023). Journalism could hold a significant competitive advantage with its long-standing commitment to fact-checking before publication.

Virtual News Media

News media should consider investing resources in developing a presence in the Metaverse while it is still in its early stages of development. The Metaverse represents an opportunity for news media to innovate and such opportunities have sometimes passed the journalism industry by and with significantly detrimental effects. During most of the past half-century, since the rise of the Internet and a host of digital media, journalism as an industry has a history seriously marred by a lack of innovation, particularly in the realm of disruptive or transformative change. The Metaverse represents an opportunity for transformative innovation in the 21st century.

Among the most notable missed opportunities for disruptive, transformative innovation in the news industry since the 1970s is the advent of digital classified advertising, namely in the form of Craigslist, and further evidenced in the advance of online advertising in digital platforms such as Google. These innovations essentially bypassed the news industry, whose core business model had been based on revenue generated by advertising sales. For most commercial news media in the analog age, advertising revenues represented about two-thirds of their funding, with the remainder coming from a variety of smaller sources, including subscriptions and user fees. With the rise of digital media, the revenue situation for most news media has been fundamentally disrupted, with advertising revenues largely lost to digital platforms. Consequently, most commercial news media have seen their revenues plummet during recent decades, and hundreds of news operations have collapsed and exited the marketplace. Those that have remained have been forced to reinvent their funding model, especially by turning to nonprofit membership models or digital subscriptions, which have been likened to digital pennies on what were analog dollars. Distribution costs for most analog news media have declined, especially as fewer newspapers need to be delivered and consumers can simply access the news online or on their mobile device. But the costs associated with creating the news product, especially original reporting, crafting a story (e.g., writing, editing, and fact-checking, as well as producing and editing photography, videography, and interactive graphics) are still high, and will not substantially decrease in the Metaverse, although some look to the use of generative AI as a possible pathway to less expensive content production. Regulatory action such as California's Journalism Preservation Act may force platforms such as Google to compensate news media for the use of their content, and this may provide an important revenue source to news media inside the Metaverse or out.

Disruptive, transformative innovation is not something unknown to journalism, however. In the first half of the 19th century, the introduction of new high-speed steam-powered rotary press technology enabled newspaper publishers to reimagine the newspaper as a platform for the general public, not just the elite. This disruptive innovation birthed the Penny Press, which not only introduced the advertising funding model but also mass media and new types of news storytelling based on stories that would appeal to a mass public. Later in the 19th century, the invention of the electric telegraph helped lead to the formation of the Associated Press as a cooperative, cost saving innovation and also impacted the news, helping introduce the inverted pyramid news story structure in which the most important facts are presented first (Carey, 1989). In the early and mid-20th century, journalism innovators created newsreels and then radio and television news.

But since these transformative innovations, the news industry has fallen back onto largely only sustaining innovations (Christensen et al., 2015). These innovations carry less risk and less potential reward. The news industry in the United States and around the world has shied away from disruptive innovation, hoping to maintain indefinitely a 19th-century business model that produced annual double-digit profit margins for many decades. Perhaps out of fear of the risk of failure or damage to its news culture, news media leaders have yet to dive deeply into the Metaverse. Yet evidence suggests the Metaverse may introduce disruptive or transformative innovation that impacts every sector of society and the world's economy. Some news media have created

AR-, VR-, and XR-based content, interactive documentaries, podcasts, and the like. Entering the Metaverse in the near term may signal an opportunity to engage in disruptive innovation and do so before some other innovator does and pre-empts such possibility for mainstream news media or journalists, and then, as with Craigslist, it may be too late. As in the early days of the digital age, the barriers to entry in the Metaverse are much lower than in the old analog age when printing press, broadcasting towers, and even newsreels required a substantial investment of resources for production and distribution of the news. Few could afford to compete, and even if they could, broadcasting in the United States and many parts of the world also required governmental approval or licensing. Even fewer could obtain this necessary governmental authorization. In the Metaverse, governmental approval and other barriers to entry may not exist, or only involve meeting technical requirements set up by corporate Metaverse platforms. Pursuing disruptive innovation is not without risk, but it also comes with possible substantial rewards. News media leaders should recognize that not pursuing opportunities for disruptive innovation such as vigorously entering the Metaverse also comes with risk in the form of missed opportunity. The question may be whether it is better to have the possibility of continued and robust success as journalism or to die gradually as an industry from a thousand small cuts.

Staking a Claim in the Metaverse: The Platforms

In 19th-century America, interest in the western territories was booming. People were moving west and a wide swath of enterprises and institutions were drawn to the profits to be made from the mining of precious metals and minerals. From the creation of the intercontinental railroad to the laying of the electric telegraph, new technology played a key role in this westward boom. Journalism was at the heart of this development. Virtually every boomtown in the west had at least one newspaper. In fact, the number of newspapers in America reached its all-time peak during this period, with the number of dailies growing from 971 in 1880 to 2,226 in 1900 in the United States (Schlesinger, 1933). Weeklies similarly grew in this period from 9,000 to 14,000. Newspapers have been in decline ever since, with digital technology further fueling the collapse in economics, number, and audience. A fundamental question for anyone concerned about the state of journalism in the 21st century is whether news media will be one of the important players to stake a claim in the Metaverse. Will journalism organizations and their leaders stake a Metaverse claim before all the valuable claims are already made? Or, will dramatic change that transforms the economic and cultural landscape of the 21st century pass the news industry by and leave it scrambling for news scraps in the Metaverse? The advance of the Metaverse may be the force that disrupts journalism in the early 21st century just as did the rise of the Internet, the World Wide Web, and online, digital communications in the late 20th century.

Leaders of the 19th-century news industry were often at the forefront of innovation and change. From the advent of the Penny Press to the introduction of albumen photographic prints and halftone printing, news publications of the 19th century introduced stunning changes to the design, coverage, and distribution of news. But in the latter decades of the 20th century and early 21st century, that spirit of innovation in journalism has declined in many news organizations, and at least partly as a result in many places journalism has dried up entirely. As of 2023, it is reported that at least 1,600 US communities are now news deserts, with no local professionally produced journalism. Journalism and the news it produces plays a vital role in a democratic society. The public has an unquestionable appetite for quality news and information about the world. As the Metaverse increasingly becomes the place where people live, work, and play, there is an opportunity for journalism to establish a substantial presence and to meet the public's need for news that is accurate and truthful. The challenge for journalism today is to not let the digital parade once again largely pass it by.

Numerous digital enterprises have proclaimed their intention to stake a claim in the Metaverse, announcing plans to build or engineer a Metaverse platform. Facebook in 2021 even renamed itself Meta Platforms Inc. to highlight its Metaverse intentions and perhaps to deflect a chorus of criticism and bad publicity the social network company had been generating. A diverse range of Metaverse platforms are open or in development. Figure 3.1 shows that as of February 2023, there are at least 44 publicly oriented Metaverse platforms; three indicate they are still in development and not yet open for use. Most (32) are blockchain based. Blockchain enables the platforms to support cryptocurrency exchange including the sales of nonfungible tokens (NFTs), virtual property, and other digital assets such as Ordinal Inscriptions, which are similar to NFTs, but inscribe the data content into "the witness of the Bitcoin transaction" (Nelson, 2023). Many of the platforms are game-oriented with cryptocurrency-incentive structures.

Some game platforms are not necessarily promoted as Metaverses. *Second Life* is an online simulation game. It is often not considered a Metaverse platform, despite having many of the qualities of a Metaverse, in that it is a virtual world, interactive, and multisensorial. It predates the current generation of Metaverse platforms and many of them are modeled after it to a certain degree. However, in contrast to many Metaverse platforms, *Second Life* is not built on blockchain and that is a main differentiator, although several

1. Africarare (Africa) Blockchain
2. Axie Infinity Blockchain
3. Black Metaverse Blockchain
4. Bloktopia Blockchain
5. CryptoVoxels Blockchain
6. Decentraland Blockchain
7. Ertha Blockchain
8. Gather.town (Korea)
9. Horizon Worlds (Meta)
10. Illuvium Blockchain
11. IoTeX Blockchain
12. MegaCryptoPolis Blockchain
13. Metahero Blockchain
14. Metacity Blockchain
15. Minecraft
16. Mines of Dalarnia Blockchain
17. My Neighbour Alice Blockchain
18. Neoworld Blockchain
19. *Nintendo Switch Animal Crossing: New Horizons*
20. Polygon Blockchain
21. Pokémon GO
22. Roblox Blockchain
23. Second Life
24. Solice Blockchain
25. Somnium Space Blockchain
26. Sorare Blockchain
27. Spatial Blockchain
28. Star Atlas Blockchain
29. Tangra
30. The Binance Smart Chain Blockchain
31. The Sandbox Blockchain
32. Treeverse (in development)
33. Ultra Blockchain
34. Upland Blockchain
35. VictoryXR
36. Virbela
37. Virtua Blockchain
38. Viverse (HTC Taiwan) Blockchain
39. VoRtex (Microsoft, in development)
40. WAX (Worldwide Asset Exchange) Blockchain
41. VXiRang (Baidu, China, in development)
42. Klaytn (open-source public) Blockchain
43. NFT Foundation (Web3 destination, NYTimes sold its NFT) Blockchain
44. Thailand Multiverse Bridge Platform (state-sponsored) Blockchain

Figure 3.1 Publicly oriented Metaverse platforms 2023.

popular platforms such as Minecraft are generally considered a Metaverse platform and also are not on blockchain. *Second Life* is one of only two platforms that have featured a news media presence. A study showed that the journalism in *Second Life* largely paralleled real-world journalism content, featuring original reporting and even calling the news products newspapers, although there is no paper involved. Brennen and dela Cerna (2010) state, "We suggest that journalism in *Second Life* focuses on community building and education, considers the influence of the on-line world to resident members' off-line lives and raises important questions about freedom of expression." It is worth noting that although this study was conducted in 2010, a decade before the contemporary growth of interest in the Metaverse, one of the three *Second Life* newspapers featured "Metaverse" in its name.

One corporate-owned platform that has captured considerable attention is Meta's *Horizon Worlds,* a self-proclaimed Metaverse that does not operate on blockchain (Lucia, Vetter & Adubofour, 2023). Launched in beta form in late 2021, *Horizon Worlds* is Meta's first version of such an immersive Metaverse (Meta Platforms, 2021). It has 200,000 active users (Clement, 2023). Although it is possible to access a 2D version of the three-dimensional (3D) environment via computer or handheld, to enter the fully immersive platform requires the user to wear a Quest headset (there are several versions; 20 million have been sold; Murray, 2023), which is a head-mounted display (HMD) with hand controllers for a haptic interface, and install the *Horizon Worlds* application on the wearable device. As with *Second Life* and *Fortnite,* the *Horizon Worlds* user creates a customized avatar through which they enter and engage VR. Hoping to make it more than simply a niche platform, Meta has designed *Horizon Worlds* as a combined social network, game environment, and workspace. With Oculus Studios producing immersive content for its VR platform, and Meta's Quest headset, it is an efficient use of existing resources. It is not surprising the company would design its approach to the Metaverse as an extension of its existing VR platform. An example of a game space developed for *Horizon Worlds* is *The Arcade* (Figure 3.2). It is a user-created place on *Horizon Worlds* where users can find a curated collection of popular games on the platform (MetaQuest, 2022). *Horizon Worlds* also features Horizon Workrooms, where users can hold virtual, or remote, business meetings. Although no news operation has set up a newsroom in *Horizon Worlds,* about a third (34%) have implemented a remote or virtual working model (Scire, 2021).

Metaverse platforms are increasingly a global phenomenon developed across the planet. Baidu has launched *XiRang,* China's first Metaverse platform (Reuters, 26 January 2022). Taiwan-based HTC has developed Viverse, a Metaverse platform optimized for its VR system Vive. HTC co-founder and Chairwoman Cher Wang explains that HTC is designing

Figure 3.2 *The Arcade* in *Horizon Worlds.*

a mobile phone optimized for AR and VR, aiming to make it a means for users to enter Viverse (Knight, 2022). South Korea's Ministry of Culture has announced plans to develop a Metaverse version of the King Sejong Institute, a government-funded Korean language education program (All News, 2023). In an effort to resist global digital giants, the French government has announced plans to create its own France-based Metaverse platforms (Park, 2023). UK-based Improbable has launched MSquared. It contrasts with the walled garden approach of Meta's *Horizon Worlds* as a network of Metaverses designed via the Metaverse Markup Language for developers.

A small number of Metaverse platforms are public or nonprofit, such as Klaytn, an open-source public blockchain Metaverse foundation. The majority of Metaverse platforms, however, are corporate-owned and controlled for-profit enterprises. Those operated by for-profit companies include Axie Infinity, Decentraland, Mirandus, Sandbox, and Virtualand. Decentraland features the Onyx Lounge, a virtual space JP Morgan owns. Sandbox's approach is to create immersive, interactive cinematic experiences, but at site-specific venues (Kelly, 2022).

Universities are among the wide spectrum of organizations creating outposts on various Metaverse platforms. Their actions demonstrate the viability of creating Metavserse experiences featuring quality, truthful content. Universities that have already opened a campus in the Metaverse, or Metaversity, include Davenport (Iowa) University, University of California at San Diego (UCSD), Thammasat University (Thailand),

Tecnológico de Monterrey, and Fundación Universitaria San Pablo. Tecnológico de Monterrey has designed a virtual campus and begun teaching courses and hosting seminars on the Virbela platform. Thammsat's Metaverse campus is built on the Thailand Multiverse Bridge Platform (TheNationThailand, 2022). Fundación Universitaria San Pablo (CEU) has designed a Metaverse learning environment via Minecraft (CEU Universities, 2021). Virbela is a Metaverse platform growing in popularity among universities for its open educational design. Many universities have launched courses or programs on the Metaverse. Among them is Arizona State University (Dandurand, 2022), which named VR journalist Nonny de la Peña to head its initiative. Other schools developing programs in the Metaverse include the University of Kansas School of Nursing, New Mexico State University, South Dakota State University, West Virginia University, University of Maryland Global Campus, Southwestern Oregon Community College, Florida A&M University, California State University, and Alabama A&M University. These initiatives underscore the growing interest in developing programs inside the Metaverse and the importance of journalism to report about the development of education in the virtual realm.

Spatial is a Metaverse platform that supports nearly all devices and modes (web, mobile, and wearable). News media could function on such a cross-device platform. Google has announced plans to revamp its YouTube platform for Web3 and the Metaverse under its new CEO Neal Mohan. The "Google-owned video streaming platform could soon have its nearly 75 billion monthly visitors witness how immersive the metaverse experience can be" (Parashar & Sharma, 2023). If Google connects YouTube to the Metaverse, this could have a powerful impact on the overall structure of the budding industry and open a pathway for journalism to deliver content to Metaverse users. Live streaming inside the Metaverse is a growing opportunity for journalism. Metaverse live streaming has been growing since AltSpaceVR live streamed an event in 2015 to users via their HMDs. Since then, a variety of events have been live streamed, including concerts, game play, and weddings (Tay, 2023).

Owned by Microsoft, Minecraft is perhaps the largest Metaverse in terms of number of users. Statista reports that Minecraft has 141 million active users worldwide (Clement, 2022). Minecraft does not support blockchain although it does support Web3 applications and NFTs (Irwin, 2022).

In 2016, *The New York Times* set up a space in Minecraft. As Thompson (2016) explains, "There's no better way to understand Minecraft than to get into the game and start exploring. Christoph Niemann, our visual columnist, worked with Hypixel, a team of professional Minecraft tinkerers based in London, to build a Minecraft world just for *The New York Times* Magazine."

The Minecraft space *The Times* created was mainly an experiment and not designed as a venue to be updated continually, although it could be.

Minecraft is gaining usage among universities for the ease with which it can be adapted and new virtual worlds created, as illustrated by *The Times* example. Analysts suggest that because of its overall popularity, malleability, and current functionality, Minecraft could prove a successful Metaverse platform generally (Pearson, 2022).

Others have pointed to the online gaming platform Roblox, in which users can build their own games, as having similar advantages as a platform for the Metaverse (Gugeux, 2022). Roblox is immensely popular especially among younger users, with a reported 42 million daily users who spent 10 billion hours inside it during the first quarter of 2021 (Mulqueen, 2022; Herrman & Browning, 2021). Created before the current generation of blockchain Metaverse platforms, Roblox generally has not used blockchain. But in 2021, it introduced its first blockchain-powered virtual world, PlayDapp Town. Creating a news virtual world might be relatively simple soon on Roblox. Taking advantage of generative AI, Roblox has announced plans to enable users to build their own virtual worlds from text entry (Hatmaker, 2023). Gamefam has announced its plans to build an immersive world for Cirque du Soleil inside Roblox. News media could use Roblox to build their own Metaverse platform.

Yet, a close examination of Roblox reveals some of the dangers of letting anyone create a virtual world: one user creation was a virtual concentration camp (Gugeux, 2022). Roblox also supports transactional services, with some users reportedly earning millions of dollars selling digital goods (Baum, 2022). Nike is among the companies that have set up virtual storefronts in the Metaverse via the Roblox platform, reportedly attracting seven million visitors from more than 200 countries (Ochanji, 2022).

Observers including Pearson and Youkhana at University College London (UCL) (1 April 2022) have noted that the Metaverse to date has failed to achieve a disruptive look and future designs need to develop a more diverse approach, and contrast with real-world structures, if they succeed in drawing users in large and sustainable numbers. For news media, this suggests they should design a Metaverse presence in a manner that does not necessarily parallel their real-world presence. Instead, they might think more creatively and design their newsroom or news structure to be more thought provoking and engaging, in a fashion that may contrast with their physical form. In the past, newspapers often built imposing real-world edifices to house their newsrooms and physical plant (e.g., printing facility) to suggest and reflect the power and authority of the press. In the Metaverse, news media should embrace a fundamentally different model and concept, one that better reflects 21st- century journalism built on diversity, equity, and inclusion.

Diversity and Inclusion

Some historically Black colleges and universities are using a platform called VictoryXR to create a Metaverse presence. Morehouse College, for instance, uses the platform to teach its students' Black history. "Students have toured the Underground Railroad, a slave ship and other artifacts using virtual reality technology," explains Morehouse (Bellamy, 2022). Clark Atlanta University received a $11.8 million grant to draw its students to a Metaverse program (Dooley, 2023). Other traditionally underserved or marginalized communities, including Native Americans, which have been particularly impacted by the digital divide, are also looking for opportunities on the Metaverse. Among the Metaverse platforms some Native communities are considering are Ertha, which is built on the Ethereum blockchain, Epik Prime, which focuses on gaming, and Metacity, which uses blockchain to support a VR cityscape (Native News Online, 2022).

Illustrating the development of the Metaverse in terms of diversity, equity, and inclusion is the Black Metaverse (2023). Self-described as "The First Metaverse Network For the Culture," its immersive design is intended to build a global community to shape the "trajectory of the creator economy." Similarly, women are also seeking to establish a presence in the Metaverse (Reader, 2022). Black Female Founders(BFF) is an online community with 14,000 members and is designed "to teach women how to get in on the crypto boom" with women-oriented NFT collections such as Boss Beauties, WomenRise, and Crypto Coven. Leading the way is Brit Morin, formerly of Google and the founder of Brit + Co. Crypto Coven sells avatars for the Metaverse. Black Women in Crypto is encouraging more black women into crypto. Meta reports that women are among the most frequent users of its Metaverse product, Spark AR Go, an AR filter production studio. "We're seeing that a really significant proportion of those filters are actually being created by women," notes Nicola Mendelsohn, VP of the global business group for Meta (Pantony, 2022).

Underscoring diversity and inclusion in the Metaverse globally is South Africa's MTN. Announcing its entry into the Metaverse in early 2022, MTN became the first African company to stake a claim in a virtual world. MTN purchased land in Africarare, which is among Africa's first VR Metaverse platforms (Monzon, 2022). Investment in the Metaverse is also developing across the global south. In India, Tech Mahindra, for instance, in 2022, launched its Metaverse practice, TechMVerse (Majumder, 2022). Likewise, Chinese computer giant Lenovo has announced plans to develop its Metaverse presence (Jennings, 2022). Virtual land is projected to increase in value. Kamin (2023) states that real estate in the Metaverse will increase in value from \$1.4 billion in 2022 to \$5.37 billion by 2026 (Kamin, 2023). However, it is worth noting that as of mid-2023, much of the early investment in the Metaverse, including virtual real estate, has seen value drop dramatically, roughly pennies on the dollars. Whether or how soon this situation may rebound is uncertain.

Technology industry leaders such as Meta CEO Mark Zuckerberg have described the Metaverse as the future of the Internet. And there is no doubt at least some utility in such a perspective, as multiple tech giants, are investing billions of dollars in developing their own version of the Metaverse (Culliford & Balu, 2021; Warren, Kalogeropoulos & Najarro, 2022). Meta has announced plans to create 10,000 Metaverse-related jobs over the next five years in the European Union, reporting billions in losses in the process (Foti, 2022; Davies, 2022). Yet such an expansive perspective of the Metaverse is somewhat vague, and perhaps intentionally so. Zuckerberg adds that his vision of the Metaverse is that it is more a time than a place (Canales, 2022). He envisions the Metaverse as a virtual realm where people come together for synchronous experiences. Yet, persistent experiences are also valuable in the Metaverse, so users can develop patterns and habits of behavior. Others have said the Metaverse is best viewed as a metaphor for virtual existence (Stein, 2022). McKinsey reports that more than \$120 billion was spent worldwide on developing Metaverse technology in the first five months of 2022 (Ball, 2022).

Zuckerberg also has argued that building the Metaverse network infrastructure (Sawyers, 2022) should be a collaborative effort. This could have the benefit of producing a standardized infrastructure that would better enable worldwide connection to and building upon the Metaverse. Different standards could work against shared development and easy access. In Zuckerberg's view, the Metaverse will not only be synchronous but also interconnected and interoperable. People would "live" and work inside and move between virtual worlds (Sawyers, 2022; Lang, 2022). Interoperability in the form of crossplay of popular games across platforms demonstrates the power of such an approach (Crossplay, 2019). Underscoring the wider interest in interoperability across the Metaverse, OpenXR is pushing to standardize advanced haptics for VR and AR (Ochanji, 2022). Conversely, a single set of standards may act to limit innovation, especially if a single or small number of organizations are in overall control of those standards. Some companies are partnering to design the Metaverse. Among these is Lego, which is partnering with Epic Games to design the Metaverse in a form that will appeal to children (Stanley, 2022). Epic developed the Unreal video game engine in 1998 and it offers a potentially powerful platform for news media to create their own game-based metaverse experiences. arcade.

studio is another platform for creating game-based metaverse experiences and does not require coding capabilities.

Ravenscroft writes in *Wired* (2021) that stating what is meant by the "Metaverse" in 2023 is akin to having a discussion about what the "Internet" meant in the 1970s. The foundation of the Internet may have been in place five decades ago. But the layers that future innovators would build upon that foundation shaped its form and function. Ensuing developments led the Internet to evolve dramatically in the decades that followed its initial invention. Despite the uncertainty, or perhaps because of it, a number of major technology giants are also developing plans to engage the Metaverse, or at least some of its underlying technologies, such as AR, VR, or other parts of XR. Among these is Apple, whose mobile and personal computing technologies have often set the standard for excellence in UX (Pandey, 2022).

Another technology giant that has taken significant strides with regard to the Metaverse is Sony. Amenabar (2022) reports that Sony is moving into the Metaverse as a gaming platform. Sony's PlayStation has introduced the PS VR2, which has been described as the next generation of the platform's VR headset. Yujin Morisawa, the senior art director at Sony Interactive Entertainment, said "the PlayStation's new VR headset is 'the hardest product' he's ever designed," states Amenabar (2022).

Similarly, Microsoft has developed the HoloLens. It blends virtual and physical worlds. The HoloLens (and HoloLens 2) is designed to dissolve the boundary between the physical and virtual worlds, creating MR, a form of XR (Khaleej Times, 2022). MR enables a variety of new possibilities including allowing a user immersed in the Metaverse to control robots in the physical world (Brodsky, 2022). MR can be used for everything from 3D games to immersive news (WooGlobe, 2023). Holography in general is emerging as an important platform for Metaverse experiences. Asmodee Digital is a unit of the French board game publisher Asmodee Group which created the hugely popular *Settlers of Catan* game. Asmodee Digital is developing holographic tabletop games that could be offered via the Metaverse (Melnick, 2022). Combining MR with haptic technology "could allow a person wearing an MR headset to use a real world object to trigger a virtual world reaction, like hitting a video game character with a real world baseball bat" (Reuters, 2022). For journalism, this suggests a blended form of interactive storytelling that engages users in multisensory fashion.

Funding Considerations

Capital is one of the primary financial components needed to develop the Metaverse. Mobile communications company Qualcomm has established a $100M fund to support the development of the Metaverse (Hayden, 2022a).

Dubbed Snapdragon Metaverse Fund, this capital is intended for companies building "unique, immersive XR experiences, as well as associated core augmented reality (AR) and related artificial intelligence (AI) technologies," Qualcomm states. Such funding is available internationally. Dubai-based Content partners funded 100 Metaverse content productions in 2022 (Vivarellii, 1 April 2022). Innovative news media could apply for funds to support the development of XR-based journalism. Illustrating the value of capital in this capacity, AR start-up Auki Labs used $13 million in seed funding to build a virtual pitch deck in the Metaverse (Wang, 2022). Similarly, with Sony backing, Tokyo-based H2L has designed an armband that uses electrical stimulation to give users low-level physical pain or mild discomfort like a bird pecking on the skin (Paleja, 2022). Such haptic interaction could be useful in a variety of Metaverse applications, from games to therapy to news experiences (Kickstarter Design & Tech, 2022). NextMind, a unit of Snap (which developed Spectacles AR glasses), has raised $4.5 million to develop a brain–computer interface for next-generation AR glasses (Heath, 2022). This would enable users in the real world to engage in an AR Metaverse experience through their thoughts. Snap is also expanding its AR lens uses to market its films in the Metaverse in partnership with Netflix, Sony, and others (Jarvey, 2022). Other companies, such as Ramen VR (developer of *Zenith: The Last City*, a crowd-funded VR MMORPG), are using investor capital to design new Metaverse platforms (Hayden, 2022b). News organizations could take a similar approach by seeking investor or crowdfunding to support models of community-engaged Metaverse immersive news reporting.

Few observers likely could have predicted in 1972 that Sir Tim Berners-Lee would in 1989 invent and launch the digital publishing platform he dubbed the World Wide Web (WWW) and that the Web would transform the shape of digital media yet to come. Likewise, few could have predicted the development of social media and all that they have wrought in the 21st century. It is similarly unlikely that observers or critics in the early 20th century could have anticipated the advance of ubiquitous mobile communications. Yet there were mobile phones on trains as early as the 1920s and there were automobile phones during World War II and engineer Martin Cooper tested the first working cell phone in 1973 (Hardy, 2022). Finally, few could have imagined the rise of global broadband wireless Internet connectivity, streaming media, or the massive disruption these and other digital developments would bring to traditional journalism and media.

Structural Imperatives

The Metaverse offers news media at least four potential organizational advantages over real-world operations. Because of the digital and networked nature of the Metaverse, news media can create structures that are (1)

Flat (or less hierarchical), (2) Flexible, (3) Agile, and (4) Collaborative. These four organizational qualities can help news media operate more efficiently, and thereby reduce costs. Moreover, these qualities enable news media to innovate in a more timely and less expensive fashion than in the physical world where the sunk costs in infrastructure can limit rapid adaptability.

One Metaverse-based organizational form that can utilize the above four structural dimensions is a decentralized autonomous organization (DAO). "A DAO is a way of organizing people and their interests on the internet using the blockchain," explains Sullivan (2022). Combining blockchain, AR, VR, and Web 3.0 is emerging as a strategy for an approach to establishing a presence in the Metaverse that spans the real world and the virtual (Konnanath, 2022). A DAO news operation could support such an approach in the Metaverse (Campa, 2022). However, a federal judge's 2023 ruling may limit the utility of DAOs by raising potential legal liability concerns (Lutz, 2023).

Pokémon Go is among the companies developing AR-based Metaverse approaches with its new game, "Peridot" (Lang, 2022). This is part of an overall trend toward ambient computing. Ambient computing means digital technologies that are nearly universally present and Internet connected, usually wirelessly. Ambient digital devices (including the Internet of Things or IoT) become almost a part of the background, with users increasingly unaware of their presence and engagement (including tracking) and are continuously in use, making the Metaverse nearly universally engaged (Rubin, 2022). For journalism, this presents an opportunity to rethink how to report the news, gather the facts, and engage sources and the public, but also challenges how to respect and protect user privacy. This involves developing more immersive content, data-driven content, and providing strong digital protections for user data within the Metaverse.

Blockchain and Crypto

The blockchain, and organizations built using it such as a DAO, could be an ideal structure for news media operating in the Metaverse. It could enable the design of secure platforms to help guard against hackers or other types of cyberattacks seeking to corrupt or otherwise adversely impact news media operations and content and possibly attempt to maliciously distribute misinformation under news operations' Metaverse brands. Moreover, a DAO would support news media revenue approaches that utilize cryptocurrencies such as Bitcoin and Ethereum and transactions including NFTs as well as protect IP rights.

Blockchain and cryptocurrency are essential for conducting transactions in the Metaverse, including NFTs, which likely will play a role in the economic future of news media operating in the Metaverse. Axie Infinity represents a convergence of games and transactional revenues within the Metaverse. Using blockchain technology, users play games to earn cryptocurrency and NFTs. Blockchain does not mean the Metaverse will prove impregnable to criminals, however. In fact, in 2022, the blockchain powering the NFT game offered by Axie Infinity was attacked. Hackers stole an estimated $600 million in cryptocurrency (Zeitchik, 2022). The second half of 2022 was not good for crypto, in general, with a massive drop in the value of cryptocurrency and the bankruptcy of FTX (Allison, 2022). Still, crypto may eventually recover somewhat and play a useful part in the future of virtual news media.

NFTs are one-of-a-kind digital objects. NFTs represent an opportunity for innovation in digital media. When an NFT is created and sold, the buyer gets a digital certificate that verifies that the buyer owns the original digital copy and a record of the transaction is recorded on the blockchain (Pogue, 2022). Since 2020, NFTs have sometimes sold for huge amounts in a variety of media-related fields, including the arts, sports, and journalism. Some art NFTs have sold for millions (Tabuchi, 2022). *The New York Times* in 2021 created the first journalism NFT. It sold for $560,000 (Roose, 2021). Also that year Gannett, the publisher of *USA Today*, created its first NFT, based on the first newspaper sent to the Moon in 1971, auctioning it off on 28 June 2021, the year marking the 50th anniversary of Apollo 14 (USAToday, 2021). The entire NFT market grew to $2.5 billion in the first half of 2021 (Reuters, 2021). NFT sales reached a record $900 million in a single month in August 2021 (Coinquora, 2021). In 2021, freelance journalist Kyle Chayka created Dirt, a newsletter funded by NFTs. In one week, Chayka earned $33,000 selling 131 NFTs and in the first 24 hours, he earned more than $20,000. For creative types, it is also possible to create and sell an NFT via AR. Using the AR drawing app SketchAR, a journalist (or any other user) can doodle or convert a photograph for sharing or selling via blockchain on the OpenSea marketplace (Kimball, 2021).

NFTs and their value can be ephemeral, making investing in them risky. Once an NFT has been created, it might disappear from public view, as was the case with the Charlie Bit My Finger NFT, removed from YouTube soon after its sale and set to unlisted (Morales, 2021). Reflecting the declining financial potential of NFTs, CNN in 2022 closed its NFT marketplace, the Vault. Also, because they use blockchain, NFTs have environmental consequences, as cryptocurrency mining has a huge energy footprint (more daily electricity consumption in 2020 than the country of Argentina).

IP Rights

IP rights are emerging as an important legal and financial issue, as revenues can derive in a substantial fashion from those rights. IP rights frequently pertain to news media and will no doubt be an important dimension of the operation of journalism in the Metaverse. Trademarks are a case in point. Mastercard has filed 15 trademark applications for its brand name as represented in the Metaverse (Nunez, 2022). Reflecting the growing popularity of VR sports including baseball, basketball, football, golf, and soccer (Carlton, 2022; Mims, 2022; Sherr, 2022; *The New York Times*, 2022; *USAToday*, 2022), NBA star LaMelo Ball trademarked his name and likeness in the Metaverse. Attorney Josh Gerben of Gerben Intellectual Property, MB1 Enterprises, LLC, "filed three trademark applications on Feb. 15, with the US Patent and Trademark Office for '**LAMELO BALL**,' '**1 OF 1**,' and '**M.E.L.O.**' The trademarks cover a wide range of virtual and physical goods and services that will allow Ball to make his presence known in the Metaverse. These goods include everything from art to apparel, trading cards to NFTs, and beyond" (Doykos, 2022). For news media reporting on sports and celebrities in the Metaverse trademarks such as these present important implications, including how to represent their likeness in the Metaverse. In addition, news media need to act soon to obtain any trademarks for their own brands and products in the Metaverse before others might take action regarding their IP rights and subsequently cause the price to climb substantially like a web address in the dot com era of the Internet. Sports such as the NBA have taken substantial strides regarding their Metaverse presence by investing in and broadcasting regularly real-world basketball games captured and shown immersively in VR, including via Canon Netaverse. Digital asset management is another major factor for news organizations operating in the Metaverse (Primrose, 2022). Meta is testing tools for digital asset management for its Meteverse platform (Reuters, 2022). Meta also has tools for its own asset management, with plans to take a 50% cut on all NFT sales on its platform (Shead, 2022).

Studies suggest that (Al-Saqaf & Edwardsson, 2019) blockchain could play a vital role in the economic fortunes of a news industry that has struggled financially in the digital age. An explorative study reveals blockchain's potential to help journalism move toward a sustainable business. Al-Saqaf and Edwardsson (2019) critically examine Civil, "a blockchain-based protocol that aims to use cryptoeconomics to incentivise the production of quality journalistic content." The study indicates that a key advantage of the Civil newsroom blockchain-based model is "the ability to enhance news credibility." The protocol leverages "decentralisation, equality, transparency and accountability," which combine to offer benefits to advertisers, news gatekeepers, and media owners.

Affordances: From AR to XR

The Metaverse is likely to feature the entire spectrum of XR. Thus, the Metaverse essentially is a media world. XR has important uses in many fields, including mental health therapy, treatment, and training. Virtual travel for the elderly or during the COVID-19 pandemic has emerged as a major application of XR. Athlete training is a widely used VR and MR application. Interactive art installations using XR have been demonstrated to be powerful in terms of user impact (Oruganti, 2022). Studies increasingly show that XR-based storytelling in journalism has a significant impact on users including in terms of various outcomes especially increased empathy (Paananen et al., 2022).

XR's blending of multisensory engagement and immersion presents unique affordances to journalism and media-related fields. Affordances are the capacities that an environment (e.g., XR) enables for the user. In his book, *The Senses Considered as Perceptual Systems*, Gibson (1966) introduced the term "affordance" in this context and subsequently elaborated upon his theory (1977). Research during the decades since has applied Gibson's theory to the arena of XR, which had not yet been developed in the 1960s and 1970s, though immersive and multisensory media forms were being developed in experimental research. Among the earliest research laboratory, VR platforms was Martin Heilig's "Sensorama" developed in 1957 featuring an arcade-style cabinet. The Sensorama enabled up to four people to experience its main affordance, "the illusion of reality using a 3-D motion picture with smell, stereo sound, vibrations of the seat, and wind in the hair" (Pimentel & Teixeira, 1993). Heilig tested his system using 3D (the essence of immersive in a media form) movies he created such as *Dune Buggy* and *Helicopter* intended to highlight motion and depth for the user's immersive and multisensory experience and to generate a sense of reality. Stereo photography (3D) is based on a much earlier invention by Sir Charles Wheatstone, who designed the Stereoscope in 1833 (Bowers, 2001). The first functional wearable multisensory system was developed in 1968 by A.M. Turing Award winner Ivan Sutherland and his student Bob Sproull (ACM Awards, 2019). Some consider theirs the first VR and AR system though neither term had yet been coined (Norman, 2023). Sutherland and Sproull's HMD was a precursor to the Metaverse UX, though the 1968 system was not networked and the Internet did not yet exist. Sutherland and Sproull's HMD was dubbed the *Sword of Damocles*, after the Greek myth.

The author's ethnographic study of *Horizon Worlds* suggests four main perceptible root affordances and several branch affordances within each on the Metaverse platform (Pavlik, 2022). The root affordances are to act, sense, interact, and create. To act is essential to the other affordances. Action begins with the user's headset and controllers and intersects with the platforms' functions that allow the user to move about, grasp, and manipulate objects. Branch affordances derive from root affordances and include taking photos, playing games, and attending events. They could include experiencing the news. To sense includes sight, sound, and touch. To interact includes speaking and listening as well as sharing experiences on social media and partying with others or playing with virtual objects, such as bouncing a ball.

Creation is a unique affordance in Metaverse platforms. In *Horizon Worlds*, users can customize their experiences, design spaces or worlds, generate symbolic meaning, and present it to others and this could take the form of immersive news experiences, including immersive documentaries created within *Horizon Worlds*. Enveloping media are psychologically powerful. Near-zero latency of multimedia sensory engagement generates a compelling experience. Unintended affordances (per Merton, 1936) such as bullying, abusing, or scamming also may occur and could be in the context of shared immersive news experiences. They could impact the user's mental health, privacy, and potential to be abused and news media need to be aware and take steps to mitigate these adverse outcomes. Platform owners have their own affordances, such as user tracking and data selling—which are largely hidden to the user—and are also important for Metaverse journalism to take into account. New affordances may be on the virtual horizon. Meta has announced it is developing an AI-enabled universal translator for its version of the Metaverse (Fried, 2022). Questions remain as to the quality or reliability of AI language translation (Johnson, 2022). Meta is also developing a voice-controlled interface for users of its Metaverse platform to navigate and simplify building in the virtual realm (Torkington, 2022). Meta also has developed and released its own AI technology, LLaMA. It is a 65 billion parameter large language model that can power chatbots, including on the Metaverse (MetaAI, 2023).

Creating content within the Metaverse represents an important and significant opportunity for journalism. Although few news media efforts have gotten underway in this regard, a variety of media productions have been shot entirely within the Metaverse, from movies to music videos. Snoop Dogg's music video "House I Built" was produced entirely inside the Metaverse platform Sandbox (Melnick, 2022).

Research suggests Metaverse XR-based user experiences could produce affordances in at least four ways pertinent to journalism. These include virtual presence, user embodiment (or embodied experience), first-person perspective, and empathetic engagement within the Metaverse. Biocca, Kim, and Choi (2001) have tested how XR may generate telepresence or user sense of virtual presence across space or time. Presence in a virtual environment can operate in a range of communication settings, including learning environments and journalism. Thereby a user feels a sense of being in a space other than their own actual physical location (Pavlik, 2022).

Embodiment in immersive and multisensory environments such as the Metaverse refers to a user's sense that their avatar has physical attributes. These may parallel the user's physical form in the real world. They may be extended or amplified for persons with disabilities or for any user to give them superhuman capabilities. Embodiment in AR and VR can generate "a realistic and particularly potent experience for the user that is more powerful than the exposure to noninteractive imagery" (Fox, Bailenson & Tricase, 2013, p. 931; Kilteni, K., Groten, R. & Slater, M., 2012; Ahn, S.J. G., Bailenson, J. N. & Park, D. 2014). Banakou, Hanumanthu, and Slater (2016) have shown that the virtual embodiment of a white person in a Black virtual body can facilitate a reduction in implicit racial bias. This represents what could be profoundly important for Metaverse journalism seeking to improve diversity, equity, and inclusion in the news. Metaverse platforms are increasingly expanding user embodiment. *Horizon Worlds* in 2022 began enabling users to see and interact with their entire virtual body (Goode, 2022). The first-person perspective of immersive media means users see, hear, or otherwise experience the virtual realm as if through their own eyes, ears, and other senses (e.g., touch). For the Metaverse, the first-person perspective makes it seem as if the user is a participant, and this is a vital consideration for journalism as it designs content to utilize this vantage point that is in contrast to the third-person viewpoint typically used in traditional journalism.

Empathy means the capacity to see or sense an experience from another person's perspective (Herrera, F., Bailenson, J., Weisz, E., Ogle, E. & Zaki, J., 2018). Studies have shown that first-person perspective experiences can enhance empathy. Investigations by Sundar, Kang, and Oprean (2017) and Wu, Cai, Luo, Liu, and Zhang (2021) have shown that AR and VR in news storytelling can be employed to enhance user presence and empathy.

Specific affordances depend on the particular Metaverse platform as well as the user interface and wearable device. Haptics is emerging as a potentially transformative UX that would enable users to feel objects virtually (Caddy & Lynch, 2022). VR headsets such as Meta's Quest, HTC's Vive Pro, and the Vuzix Shield Smart Glasses each highlights somewhat varying affordances. The Quest, for example, features haptic control, "mixed reality," and face-tracking capabilities and supports physically active experiences such as exercise (Hunter, 2022). The Vuzix provides hands-free smartphone integration (Hacking, 2022). The Quest is the market leader. Quest 2 captured 78% share of the combined AR/VR marketing 2021, according to a report by research firm IDC (Patra, 2022). China's ByteDance (parent of TikTok) is quickly gaining on Meta in the VR headset market with its Pico system climbing to number two in global market sales (Barr, 2022). Wurmser (2022) reports that "Once the domain of gamers and young social media users, AR and VR are entering the mainstream." Introduced in 2023, Apple's Vision Pro MR headset, with eye, voice, and gesture control over content overlaid onto the real world, may fundamentally impact both the marketplace for AR and VR and the development of the Metaverse. Apple launched its headset on 2 February 2024, and preorders are said to top 180,000 units (Lai, 2024).

The mainstreaming of the AR and VR consumer market reflects the continuing improvement in the design and dsets and AR technology. It is expected AR users will continue to outpace VR users. As such, many consider AR to be the most effective pathway to the Metaverse (Mingay, 2022). There are signs Meta is shifting its Metaverse focus away from VR to AR (Bajarin, 2023). Meta's corporate interests in advancing its own stand-alone VR HMD has in many ways limited the development of quality VR experiences and thereby the Metaverse itself. Stand-alone devices such as the Quest Pro are technically inferior to what could be made available in VR systems tethered to high-powered computers. But the lower processing capacity of current stand-alone systems limits the possibilities of immersive content design. Even with its $1500 price, the Quest Pro and similar stand-alone VR headsets struggle to run intensely immersive PCVR apps such as Google Earth VR.

It is worth noting that the US Department of Defense, which funded the ARPANET and led to the development of the Internet, has begun utilizing the Metaverse for its meetings and training programs, and has employed the Oculus (Meta) VR platform in this regard (Eversden, 2022). Meta also has announced plans to develop two types of AR smart glasses, one for the high-end and one for the low-end market by 2024 (Statt, 2022).

Future Economics

As much as anything, the development of the Metaverse is about economics. The Metaverse is projected to serve as a major engine of economic growth. During the World Economic Forum in Davos, Switzerland, in 2023, analysts turned much of their attention to the Metaverse, gaming, and Web3. "CoinDesk's Market Index shows that the Culture and Entertainment sector, which contains media, content, and gaming projects, is up roughly 46%

in 2023" (Yahoo!finance, 2023a). This projected economic growth is a major reason for the rising interest in developing and investing in the Metaverse, which emphasizes gaming and Web3 applications. With media as an important part of the forecast, there is an opportunity for the journalism industry to invest in the Metaverse and see potential revenue returns though not without financial risk. In 2024, Davos used Microsoft's Mesh Metaverse platform to enable virtual access to its meetings (King, 2024).

Additional economic data point to the rising fortunes of investing in the Metaverse. A wide spectrum of technology companies are pivoting deeper into AI, AR, and VR (yahoo!finance, 2023b). Meta hemorrhaged financially in 2022 after its multi-billion dollar investment in VR and its new Metaverse initiative, while advertising revenues and users declined. But the company saw its stock rebound in early 2023. Meta stock is "showing glimmers of hope this year after falling 65% last year." In early 2023, Meta stock roses "25% plus," reports analyst Jared Blikre (yahoo!finance, 2023b). By March, however, Meta announced layoffs of more than 20,000 employees, a belt-tightening move like many others in the tech sector that may hint at troubled waters ahead, including its commitment to the development of the Metaverse.

Deloitte, however, is optimistic about the potential economic windfall the Metaverse may deliver. In a 2022 report on the Metaverse, Deloitte declared the Metaverse is "No longer science fiction." The report notes that the Metaverse could introduce $1.4 trillion a year into Asia's GDP (Tong, 2022). Deloitte adds that the Metaverse is likely to disrupt the streaming media industry in Europe and elsewhere (Priestley, 2022).

Conclusions

Rosenberg (2022) argues that despite some doubters who contend the Metaverse will fade, it is actually inevitable. Rosenberg is the CEO of Unanimous AI, the Chief Scientist of the Responsible Metaverse Alliance (RMA), and the Global Technology Advisor to the XR Safety Initiative (XRSI). He offers this explanation as to why the Metaverse is inevitable. "It's in our DNA. The human organism evolved to understand our world through first-person experiences in spatial environments. It's how we interact and explore. It's how we store memories and build mental models."

For journalism in the Metaverse, Rosenberg's contention suggests users will employ their natural human abilities to perceive, interact, and explore when they engage or experience immersive news content in first-person formats. For journalists and news media, it will be vital to create content that engages those senses in a fashion that supports first-person experiences in spatial environments. This will meet the public's appetite to actively explore and interact and thereby build mental models based on accurate and truthful news and information. News media should embrace this opportunity by designing their own organization and system of Metaverse journalism that engages the structure and parameters of evolving Metaverse platforms. This opportunity will be shaped by economic, cultural, and legal or regulatory frameworks both in the United States and around the world. News media need to stake a claim in the Metaverse before it fully crystallizes. By staking their claim early, news media can operate efficiently and effectively to engage a global public inside the Metaverse. The combination can help yield a sustainable and valuable news industry inside the Metaverse.

Chapter 4

METAVERSE REPORTING: LOOKING INSIDE AND OUT

In a virtual or real world, journalism begins with reporting or the gathering of information that serves as the foundation for stories or other news content. In the Metaverse, the nature of reporting shifts in potentially profound ways from the traditions and methods of real-world journalism. For example, there is no world outside the real world, but with regard to the Metaverse, we all exist outside it, unless and until we enter that Metaverse via some digital platform. Reporting about the Metaverse therefore can be conducted from either within or from outside the Metaverse. From outside the Metaverse, that reporting might look at any of a wide spectrum of phenomena, including those who own or would own, build, and control (e.g., regulate) the Metaverse as well as the real-world consequences of the Metaverse (e.g., the interconnections between social media inside and outside the Metaverse, and psychological impact of Metaverse experiences on users inside and after they exit the Metaverse).

Metaverse journalism may parallel some aspects of real-world news gathering, including first-hand observations of events and activities within the Metaverse, interviews with those people, digital twins, or AI-enabled characters who populate the Metaverse, as well as observations of the Metaverse itself especially in terms of its design, construction, and function (i.e., this is somewhat parallel to how a journalist might report about a city).

Although this chapter touches on aspects of reporting outside the Metaverse, the primary focus here is on reporting inside the Metaverse. Five factors that are particularly pertinent to the nature of the Metaverse most shape or characterize reporting inside the Metaverse. These factors are immersion, data, interactivity, multisensory communication, and AI. These factors are particularly salient to Metaverse reporting because they are essential to life inside the Metaverse and to a fundamental transformation of storytelling in Metaverse journalism.

Immersion in the Metaverse refers to the fact that users, including reporters, are enveloped in a 3D virtual world where the physical rules of the real world do not necessarily apply. Instead, the affordances of the platform determine the possibilities. Immersion shapes the context of news gathering inside the Metaverse. To do their reporting, journalists must create an avatar and enter into and journey through the Metaverse as that avatar. Reporters become participants in the Metaverse and see, hear, and otherwise experience it largely from a first-person perspective.

Data are the building blocks of news gathering in the Metaverse. Reporters collect facts, whether inside or outside the Metaverse. Inside, those facts are constituted of data. Data are the enabling essence of fact gathering in the Metaverse. Data provide the core metrics that reporters can use to describe and explain news inside the Metaverse. Metrics about virtually everything inside the Metaverse are possible. Journalism inside the Metaverse must utilize such data-driven measures to make sense of the overall patterns, trends, and developments as well as to situate their other news gathering in a wider context. The speed and scope of the Metaverse will make efficient and thorough data collection and analysis essential to quality Metaverse journalism.

How data and their corresponding code manifest in the Metaverse is a vital matter to news gathering. Inside the Metaverse, reporters have access to data in several forms. These include the numbers that quantify the Metaverse, its users, and their experiences within it. These data take a variety of forms from usage and user statistics to metrics of the Metaverse itself. Data range from the individual user level to the organizational or structural level. For example, data are available to reporters in terms of the number of users attending a virtual event, the number of likes, or user preferences. Places, transactions, and activity of various types, NFT and land sales, and other indicators of economic activity are all part of what journalists will collect data about. These data can help reporters assess news value and shape what media present to the public.

Interactivity and multisensory communications are two other essential forms in which reporters will collect data. Interaction will be in terms of engagement with other users and with virtual objects and entities. Reporters can observe and also record these interactions (e.g., a conversation between avatars) to incorporate them or re-present them in stories or other forms of news content.

Multisensory communications refer to the sites, sounds, and haptics of the Metaverse as users experience it. Reporters can capture or record multisensory media and use that material to tell their stories. While inside the Metaverse reporters can capture or record images (analogs to photographs), video and animation, audio, and potentially even haptic experience as elements to use in stories. It is also useful for reporters to add data or draw upon data to provide a wider context and to frame stories and link them to the world outside. In capturing user data, reporters will need to obtain user permission. And it will be vital to make reporting processes and methods transparent. The combination of all these methods of news gathering will enable journalists to report about the entire realm of the Metaverse, from cultural and social life to business and economic activity and the problems of the Metaverse, including abuse, human trafficking, sexual assault and harassment, or any other criminal activity that may occur. Covering culture in the Metaverse requires a global approach. Young and Stevens (2023) report, "Other places want to venture into the Metaverse, but to be successful, you need to have good content. In Korea, that content is K-pop." Korea's Kakao Entertainment is partnering with a mobile gaming company, Netmarble, "to develop a K-pop band called Mave that exists only in cyberspace, where its four artificial members will interact with real-life fans around the world." Moreover, "Kakao is also behind 'Girl's Re:verse,' a K-pop-in-the-metaverse show, whose debut episode on streaming platforms this month was viewed more than a times in three days."

A compelling domain in which journalism can put these capacities to effective use is in the creation of data-enabled 3D scene reconstruction of news events (Han et al., 2021). Using actual data from real-world news stories, such as crime scenes and other venues, 3D reconstructions could be provided inside the Metaverse for users to explore as an eyewitness or crime scene investigator with the facts assembled as a form of interactive and immersive journalism. *The New York Times* produced something along these lines using AR to allow users to explore virtually the cave in Thailand where the boys soccer team had been trapped and then rescued in 2018 (NYTimes, 2018).

The Times has invested resources in developing 3D scene reconstruction as a way to "bear witness to history as faithfully as possible. This is especially true for photo and video journalists, who put themselves close to the action in order to visually document events." A team of eight journalists at *The Times* used environmental photogrammetry to present an immersive and interactive scene reconstructed faithfully and presented in 3D (Boonyapanachoti, Cohrs, Kim, Koppel, McKeague, Porter, Rambelli & Wilhelms, 2020). The scene is an actual loft in Providence, RI, and users glide through the scene in a 3D 360-degree high-resolution visual display. Annotations are layered into the space to advance the storytelling and users can zoom in and explore in their own fashion. A graphical map of the loft is overlaid in translucent fashion allowing the user to visualize the entire scene and see where they are within the loft. "Photogrammetry is a process that involves taking dozens or hundreds of still photographs and using software to stitch them together in a 3D mesh," *The Times'* team reports (2020). By placing such 3D scenes inside the Metaverse, users could have a fully immersive, interactive, and multisensory news experience all based on precisely constructed scenes. Such 3D exploration could be a powerful way to present a wide spectrum of stories, from art to science to crime.

A growing amount of research is underway to develop 3D scene reconstruction of crime scenes for forensic purposes, and this work could be directly applied to journalism. This research indicates that "the implementation of light detection and ranging (LiDAR) scanners and immersive technologies, alongside traditional methods, has been beneficial in the recreation of crime scenes," state Maneli and Isafiade (2022). LIDAR is increasingly available at a relatively low cost in contemporary mobile devices such as the iPad. Investigations of 3D scene reconstruction indicate that the application to the Metaverse is likely via AR, MR, and VR. Maneli and Isafiade's (2022) meta-analysis of the research literature found that "20.2% of the articles implemented immersive technologies in crime scene reconstruction," of which most (75%) were VR, followed by AR (15.3%), MR (5.9%), and a VR/AR blend (3.8%).

On the horizon is the development of Neural Radiance Field (NeRF) technology. NeRF enables the creation of "novel views of a scene by modeling the volumetric (3D) scene function through a neural network" (Gosthipaty & Raha, 2021). Neural networks are the same deep learning used in generative AI. A collection of correctly positioned cameras enables the recording, rendering, and streaming of a fully navigable VR experience of live events, which could be presented inside the Metaverse as a new form of precisely reported visual journalism of breaking news. Google is working on such image and video rendering for much of the US urban landscape. 3D mapping of spaces would enable powerful news storytelling in the Metaverse merging the physical and virtual worlds.

Illustrating the potential for interactive and multisensory reporting (non-3D) is the HBO documentary *We Met in VR* (HBO, 2022)."The film follows Jenny, an American Sign Language (ASL) teacher, dedicated to building a welcoming community for deaf and hard of hearing VR Chat users," HBO states. The Sundance Film Festival 2022 Documentary Competition featured the production. *We Met in VR* was filmed entirely inside a Metaverse platform called VR Chat using Joe "Hunting's Valve Index VR headset and custom camera called VRC lens—a third-party program that allowed Hunting to create realistic shots as if this were a documentary set in the real world" (Hector, 2022; IMDB, 2022). Advances in AI are bringing the capacity to not only shoot a documentary but also to plan, storyboard, shoot, and edit an entire film, documentary, or news video inside the Metaverse (Brown, 2022).

Projects like *We Met in VR* suggest the potential for news media operating inside the Metaverse to develop community-led news reporting projects. Such community-led reporting projects have proven increasingly successful in real-world journalism. They are especially effective in working with groups historically marginalized and given minimal voices in mainstream news media (Hamada, 2023).

Covering courts is a staple of real-world journalism. It is becoming a staple of Metaverse reporting, as well, as courts from China to Colombia start holding legal proceedings inside the Metaverse. In February 2023, a Bogota municipal court held a legal trial inside *Horizon Workrooms* (Di Salvo, 2023), which enables groups to meet and conduct business in a virtual workspace. News organizations such as Reuters (2023) covered the Colombia legal proceedings, including photos captured from inside the virtual courtroom.

Metaverse journalism took preliminary form in a 2022 exhibition on the Cathedral of Notre Dame told via AR. In 2019, fire gutted the centuries-old cathedral of Notre Dame in Paris, France. During the reconstruction of the historic building, AR has offered an innovative digital medium through which to tell the cathedral's story in an immersive, interactive, and multisensory fashion. On 20 November 2022, the author was in New Orleans, LA, for a conference and had the opportunity to experience a new AR-based exhibition on the Cathedral of Notre Dame, its history, cultural significance, the fire, and its aftermath. *Notre-Dame de Paris: Augmented Exhibition* was installed as a traveling exhibition at the Historic New Orleans Collection in the French Quarter. Available free to visitors, the exhibition is scheduled to travel to New York City, Shanghai, and the World's Fair in Dubai. The user engages the AR experience via a specially crafted, dedicated, handheld tablet computer called the Histopad. Based on the author's immersive experience, the AR Exhibition is extraordinary and signals a possible form of nonfiction content, including journalism in the Metaverse through an AR portal. The Notre Dame AR experience is thoroughly researched, brilliantly designed, highly interactive, and a beautiful means to engage a powerful cultural and architectural wonder through digital immersion. To start the journey, the user scans a code for a specific station in the exhibition, which then unveils an interactive, enveloping presentation about an aspect of Notre-Dame de Paris. Images on the Histopad are in 3D, immersive, and navigable. The user can touch the objects displayed, rotate them, or zoom in and explore, hearing contextualized audio and controlling the display through an interactive timeline. White virtual buttons populate the videos and the user can touch them to access additional information and navigate the virtual space overlaid on the room in which the user stands or sits. Included are artfully and accurately designed digital animations depicting the past such as the stonework in the cathedral's construction. Historically authentic audio populates the experience from the sounds of construction to cathedral bells chiming during a wedding. Users can enter 3D spaces and explore everything from detailed architectural designs such as flying buttresses to the building's materials including wooden timbers employed in Notre Dame. Scanning additional stations continues the user's virtual AR journey. Layered into the experience is a virtual scavenger hunt in which the user searches for hidden stained glass windows actually found in Notre-Dame de Paris. The user learns about the architectural wonder, its cultural context, its role in the religious life of Paris, and historical figures from Paris's past such as the Empress Josephine, Emperor Napoleon, and novelist Victor Hugo. Similarly well-researched and designed journalism in the Metaverse would make for a highly compelling user experience.

The Special Role of AI and Reporting in the Metaverse

AI will fuel and shape the reporting process and storytelling inside the Metaverse (Nicholls, 2022). This includes generative AI but extends much further and is particularly fundamental to understanding the nature, character, potential, and pitfalls of reporting inside the Metaverse. AI has emerged as a fundamental force shaping the Metaverse, how it functions, its nature and parameters, and provides a foundation for reporting inside the Metaverse.

The Metaverse exists as computer code presented to its users as sights, sounds, and haptics enabled by varying levels of interactivity, all increasingly controlled by sophisticated algorithms, or AI. In the future, other senses may be engaged, including smell, taste, or, via BCI, new senses or an extension of existing senses that could let people sense visual maps of data or see ultraviolet light. AI likewise drives and shapes journalism within the Metaverse and the reporting it relies upon. The following discussion highlights the evolving nature of AI and how it may set the course for news gathering and journalism inside the Metaverse.

AI means machines or computers that have the ability to think or act in a fashion that mimics human thought. This means machines that can analyze data, that can make decisions, perceive images, write stories, and the like. In *The Rise of the Robots*, computer scientist Martin Ford (2015) argues that robots can replace human workers in any field that involves data and predictable patterns that can be analyzed, interpreted, and expressed. Radiologists are among the first professional fields in which robots, virtual ones, can perform with greater precision and speed than human specialists. In analyzing patient x-rays, AI can reliably spot cancer or other defects in the images. At this point, such AI tools are used to assist human radiologists but in the future, such AI applications could offer a potential cost saving in medical care without sacrificing quality. The same pattern may apply to the future of AI in Metaverse news gathering and journalism in general.

The advance of AI is driving the rise of robots, whether physical or virtual. Journalists, especially the function of news gathering or reporting, are being impacted by AI and virtual robots. This is evident in news content that is increasingly being shaped by AI, both inside and outside the Metaverse. Automated news reporting (and writing and storytelling) is already a significant part of modern news media. From the Associated Press (AP) to the *LA Times*, algorithms that gather news and information and then process it into stories for the public are happening with increasing frequency across the globe from China to the United States, and will similarly play a primary role in news gathering in the Metaverse. The AP uses such tools to report and write thousands of business and other stories every year. These examples are only just the beginning of automated AI-based systems in journalism and illustrate the potential inside the Metaverse. AI will play a vital role in areas of news in which reporting and writing are amenable to known processes (e.g., the five Ws of news, who, what, when, where, and why) and data, such as interviews and quotes from sources, business or sports statistics, or images and video of news events, especially in structured spaces such as the Metaverse.

IoT means digital devices that are connected to the Internet, either outside or inside the Metaverse. IoT has grown dramatically in number and now far exceeds humans with Internet access worldwide. It is estimated that there are at least 14 billion IoT in 2022; in contrast to the roughly 5 billion persons globally with Internet access (Hasan, 2022). IoT ranges from cameras to smart speakers. IoT increasingly has sensors that can capture a host of data. IoT also increasingly has integrated AI that allows for advanced processing of the data collected. These devices (physical in the real world, virtual in the Metaverse) are essentially functioning on the Cloud and feature virtually real-time AI data processing. AI features several areas of application, each of which may impact journalism, particularly journalism that utilizes data collected via IoT. Four of the most widely developed AI applications are as follows:

1. **Machine vision**, which means computers that understand the position or nature of objects in the world (physical or virtual) through light and cameras. Machine hearing is also advancing rapidly as voice recognition and real-time language translation develop.
2. **Machine learning (ML)** refers to teaching computers about the world through training examples or by modeling human actions or decisions. Neural networks and deep learning are prime examples and have played a key role in the development of generative AI systems.

3. **Natural language processing (NLP)** means computers making sense of human languages or digital devices that can read, write, listen, or speak to communicate with people. Smart speakers (or smart media generally) perform this function. Combined with ML, these devices are impacting the real world and the Metaverse in the form of various generative AI platforms for text (e.g., ChatGPT), images (DALL-E 3), audio, including voices and music (e.g., Jukebox), or video (e.g., Gen-1). Meta's AI platform is being made available at no cost for anyone, including users, developers, and journalists whether operating in the Metaverse or beyond. Google's commercial AI tool Genesis can write news stories, whether in the Metaverse.

4. **Robotics** refers to automation via machines that can use any of these applications to act in the physical or digital, virtual realm in place of or as a supplement to humans, including reporters. Robotics has been long in development. In the 19th century, the followers of Ned Ludd opposed the development of textile machines that automated weaving work and threatened to displace human workers (Byrne, 2013). Those followers came to be called Luddites, and this term is still used in the 21st century as a pejorative name for someone who is opposed to new technology, including robotics or other areas of AI or smart media, including those that are impacting journalism and news reporting in or outside the Metaverse. All four of these AI applications are relevant to the transformation of journalism in the Metaverse and beyond.

Smart media have emerged in the 21st century as an increasingly significant and widely available platform for journalism and news gathering. From smartphones to smart speakers to smart TVs, a large portion of the population in most developed nations has access to such devices. Per Table 4.1, Statista data show that smartphone adoption is high in virtually every part of the world as of 2021 and will be roughly 80% or higher nearly everywhere by 2025.

Moreover, smartphones are getting smarter (Holland, 2022). They have higher quality cameras and new sensors that can track or measure many more things, such as size and location via LiDAR. Smartphones will increasingly feature machine learning and act as a gateway into the Metaverse, as is suggested by HTC's AR- and VR-optimized smartphone designed as a Metaverse access device. Reporters will likewise use these smartphones to enter the Metaverse and conduct their reporting. Smartphones will include real-time language translation, and the ability to capture and display location-based AR, and more. Smartphones with AI capability will be able to assist journalists in news gathering or processing that reporting, such as generating real-time transcripts. Apps already can effectively use

Table 4.1 Smartphone Usage Rate

Region	2021	2025
Sub-Saharan Africa	64%	75%
North America	82%	85%
MENA	79%	82%
Latin America	77%	82%
Europe	79%	83%
CIS	79%	86%
Greater China	77%	89%
Asia Pacific	74%	84%

Source: https://www.statista.com/statistics/1258906/worldwide-smartphone-adoption-rate-telecommunication-by-region/

the smartphone camera to see a plant or flower and recognize it and tell the user the identity of that plant. For journalists, next-generation AI-based apps may be able to process large volumes of data or information and assist reporters in identifying patterns in those data. This might be particularly useful in investigative reporting where identifying patterns in transactional records, banking or other financial information, and the like could be useful in developing a story about corruption in governmental spending or business or surveillance activity especially in the Metaverse.

Extra-smart smartphones will be able to model the behaviors of the people who use them, including journalists. The beginning of this is apparent in smartphones that can observe user patterns and make recommendations based on those patterns. For journalists, this might take the form of an app being able to recommend new leads, stories, or even pose questions in a developing story and check facts, or an app that could identify errors in reporting across sources of breaking news, or an app that might identify plagiarism in news reports from various sources.

Advances in smartphones also will feature greater accessibility. This means improving capabilities for those persons with disabilities, including reporters. For example, smartphones will provide live captioning. This is a capacity journalists or news consumers could find useful, such as during a press conference or even an interview with a source inside the Metaverse or outside it. At a conference on education in January 2023 in Honolulu, HA, the author made a presentation about the implications of the Metaverse for higher education. Using voice recognition and real-time speech to text conversion, the author's speech was displayed on screen for the audience in real time along with his slides, making his remarks more fully accessible.

Other smart media are also becoming increasingly adopted by the public both in the real world and inside the Metaverse. This adoption brings implications for journalism engagement, including citizen reporting. Edison and NPR research data show that as of 2020 some 60 million people in the United States have at least one smart speaker, including the Amazon Echo with Alexa, Google Home, or Apple Homepod. And there are at least 157 million smart speakers in US households (Richter, 2020). Coupled with Siri and other voice-based interfaces on smartphones, the potential for smart-media engagement is becoming almost ubiquitous in the United States and much of the developed world. The Echo with Alexa can recognize each user's voice and interact in a manner customized or personalized to each user. These capabilities mean journalism can develop smart news.

Smart news can offer a wide spectrum of functions, including customization. For journalism, it is an opportunity to provide personalized and interactive audio news for individual users including citizen reporters inside the Metaverse. For instance, a news platform could provide each user with a news-gathering conversation (a virtual assignment editor) available on command from the user. Users could be provided on-demand audio news briefings customized to their interests and provide audio-based data input of their own observations inside the Metaverse, as well as data observed via machine learning, and personalized via voice command or user query.

Moreover, smart media are increasingly more than just audio-based. The Echo Show, for instance, features a high-resolution display with built in a camera that can scan the room for movement and track the user's location. The Show features a robotic platform that allows the device to move in a manner synchronized with the user's location. For journalism, these capacities represent opportunities to engage news consumers in new ways including for citizen reporters. For instance, a news avatar could be displayed on the screen and could interact via voice with the user, answering news-related questions in real time. As such multisensorial smart devices continue to develop (per Moore's Law), they are likely to serve as vital pathways into the Metaverse where a user can enter briefly or for an extended visit, take an action, or act as a citizen reporter, all via voice, glance, gesture, or touch. Recent events suggest how citizens are inclined to act as real-time reporters during breaking news events. Such was the case in Gaziantep, the epicenter of the 2023 earthquake in Turkey and Syria. Citizens shared news of the natural disaster before dawn by cellphone (Gures and Yoon, 2023).

Smart avatar interaction could be developed by using chatbots that can act as virtual news-gathering assistants. A Google chatbot even has been described by one of its engineering developers as having shown signs of becoming sentient or possessing self-awareness. Although many doubt this capacity, it is clear that chatbots are increasingly sophisticated and demonstrate a capacity to carry on a conversation with a human user. Connected to the internet via the Cloud, a news chatbot could converse with news consumers and provide the news and potentially serve as a smart platform to identify false or fake news and misinformation and conduct verified news gathering (Sample, 2022).

Immersive smart media are also poised to impact journalism content and news gathering in the Metaverse. Platforms such as the Quest, Vive, and Pico are among the most widely available in terms of consumer adoption worldwide. These systems are the HMDs of Meta, HTC, and ByteDance and they are portals to the realm of VR and the Metaverse. News media are increasingly creating immersive journalism content that could populate these virtual environments. News consumers can enter these platforms to experience the news as if present in a story. Why news consumers would want this type of immersive news engagement parallels the growing demand for participatory media such as video games and news games, especially among younger audiences. Users could also serve as citizen reporters inside the Metaverse. Likewise, these news consumers would be inclined to serve as citizen MV reporters as a further opportunity to participate in content creation and sharing, much as they participate actively in Internet-based social media. Yet MV citizen reporters would not differ significantly from other citizen reporters in that they are active users of digital media; MV citizen reporting would differ largely in that it will be more immersive.

Journalists are among those who are likely users of smart media. *The New York Times* reports (December 2019) the results of a self-study of the extent to which its own journalists are using smart media technology. Says *The Times*, "After three years and more than 130 columns, the smartphone" came out on top with almost every journalist reporting they use a smartphone. But, usage is not 100%. "There were also some deliberate Luddites among us." Yet, said *The Times*, the usage of the smartphone as a reporting tool is widespread and for varied uses. "Many reporters relied on smartphones for recording interviews and turned to A.I.-powered apps like Trint and Rev to automatically transcribe interviews into notes." Also, "Most *Times* reporters now also rely on some form of encrypted communication, particularly messaging apps like Signal and WhatsApp [...] to keep their sources and conversations confidential." As journalists enter the Metaverse, virtual smart media devices will likely play a significant role in news gathering and secure communication with colleagues, editors, sources, and the public, not to mention AI agents.

Smartphones are also useful in cultural reporting. *The Times* internal study shows, "In the world of dining, digital photography and platforms like

Instagram have become the main method that restaurants use to communicate with patrons." Reporters also curate movies and TV shows increasingly streamed onto mobile devices. Science journalism is also affected by the rise of smartphones. "Rocket launches are now live-streamed online, which let our space reporter watch from his phone instead of heading to the space station" (Chen, 2019; ProjectTopics, 2021). As such, virtual smart media will likely play an important role in reporting on culture, science, and beyond inside the Metaverse.

IoTs are increasingly smart media themselves and especially relevant to how journalists do their work both in the real world and the virtual. Among the most significant applications is facial recognition. As a form of machine vision, facial recognition depends on several elements of smart media. These include the visual data collected and its resolution and frame rate, as well as the AI technology, which can be utilized either via the Cloud or on individual digital devices available to the news media. In Metaverse journalism, facial recognition will play a role in interaction with avatars as news sources. Ensuring privacy protections while effectively utilizing this technology will be an important challenge for Metaverse journalism.

The virtual newsroom is emerging fueled by AI. Metaverse newsrooms will not be dependent on a central physical location, although they will likely still have a persistent location in a virtual world, where events might be held for live streaming. The virtual newsroom is more decentralized, efficient, and cost-effective. The arrival of Web 3.0, including transactional technologies, may further fuel the advance of the virtual newsroom especially as an interface between virtual and real-world journalists.

Public engagement has increased with the rise of smart media. Pew Research data show that as of 2021 more than eight-in-ten Americans get news from digital devices and many of these are their smartphones (Pew, 2021). This trend may well extend into the Metaverse. Citizen reporting also has surged with the rise of smart devices and portends the advent of citizen reporting via virtual smart media.

Public perception and trust are vital to the future role of smart media in news. Research (Kieslich, Došenović, Starke, Lünich & Marcinkowski, 2021) indicates that at least in Germany, the use of AI in journalistic newsrooms is viewed negatively. "Not only is there little to no presumed improvement with regards to overall journalistic quality, many citizens are also in favor of strong regulations for AI systems in media and journalism." Technology enterprises already have developed creator tools to build conversational AI inside the Metaverse (Takahashi, 2023). Los Angeles-based MeetKai has at least 50 million global users. At CES 2023, "The company unveiled MeetKai Reality, MeetKai Metaverse Editor and MeetKai Cloud AI to help people create seamless metaverse experiences" (Takahashi, 2023). Using these tools, an innovative news organization and journalist could design and launch a conversation with an AI journalist or AI editor who could act as a virtual assignment editor for aspiring citizen reporters inside the Metaverse.

Problems have emerged because of smart media. Whether citizen reporters or not, users are justifiably concerned about how smart media collect data about their activities, including what they do inside the Metaverse, and the tracking of their virtual and physical movement. Metaverse users physically often move extensively, research shows, especially with haptic interfaces. Users are rightly concerned about whether these data are being sold for marketing or surveillance purposes, and news media who may be engaging with virtual citizen reporters need to take these concerns into account and protect their users. The events of early 2023 involving a Chinese surveillance balloon over the US underscores the likelihood that such spying could happen by state actors (including Russia and China; Eady, Paskhalis, Zilinsky, et al., 2023) collecting data from the Metaverse (Sanger, 2023). Alleged uses of TikTok for similar purposes add to this concern and suggests the important role that independent journalism inside the Metaverse could play in reporting about such state surveillance. Multiple governments from the UK to the United States have banned TikTok on government-issued devices; potentially such regulatory action against Metaverse platforms could affect their development (Hassan, Mellen & Taylor, 2023). News media that utilize smart media will need strong, clear, and transparent policies that place a premium on ethics if there is to be public trust in journalism based on smart media (Tatum, 2023). Other ethics concerns are important as well. These include the potential bias that has been shown to exist in facial recognition technologies, including in the Metaverse. Studies show that facial recognition technology is biased in terms of both race and gender. One governmental study found that "Facial-recognition systems misidentified people of color more often than white people. [...] Asian and African American people were up to 100 times more likely to be misidentified than white men, depending on the particular algorithm and type of search. Native Americans had the highest false-positive rate of all ethnicities, according to the study, which found that systems varied widely in their accuracy" (Grother, 2019).

Although comparable studies of facial recognition bias inside the Metaverse need to be conducted, news media that might utilize avatar recognition technology will need strong ethical frameworks with regard to issues of social and racial justice and gender equality. Avatars often are designed to reflect their human counterpart's race, gender (gender identity), or ethnicity and the same bias that occurs in real-world use of facial recognition could occur if avatar recognition tools are employed inside the Metaverse.

With the rise of increasingly powerful AI platforms that energize these smart media, a fundamental question news media must consider is whether the next generation of smart media is a threat or an aid to better news gathering and thereby journalism. Will smart media replace human journalists inside or outside the Metaverse, or will smart journalists find ways to effectively utilize smart media to create more accurate, complete, and contextualized news content?

Conclusions: DINE in the Metaverse

This chapter has suggested that news gathering in the Metaverse revolves around five primary dimensions: immersion, data, interactivity, multisensory communication, and AI. Blending these five dimensions of reporting will enable a new kind of Metaverse journalism. This is the dynamic immersive news experience (DINE). Using the capabilities of the Metaverse, especially AI and data, DINE can function in virtual real time (i.e., with near-zero latency) or asynchronously and be customized to user preferences, interests, and profile, featuring interactive, multisensory content experienced alone or in a multiuser space. This design parallels the popular immersive asynchronous multiplayer game, *Eldon Ring* (Greenwald, 2022). Moreover, DINE will be based on actual data ensuring precision, accuracy, and the highest potential for truth (Laws, 2019). To illustrate, one might consider a DINE formatted story about climate change. It could be based on the user's actual physical location (or a choice of location) and experience inside the Metaverse. A DINE story could allow the user to select a date in time, past or future, and see, hear, and feel the effects of climate change in their community (or place of choice), as if they are present in that time and space. DINE could enable precise, data-driven reporting that situates the user virtually and draws upon their social network or community in the Metaverse, inviting other users to join in the experience and together fight to reduce their carbon footprint, and then realize the benefits of change, or the adverse consequences of failing to make the needed change in time. Community members could participate in an immersive discussion about the DINE experience and how to bring the lessons from the Metaverse to the real world.

DINE contrasts with the traditional news storytelling model in which content is fixed in published form and not dynamic. The transition to the new DINE model of news already is underway, although very gradually in the digital networked age. But it likely will evolve to a much more advanced level in the Metaverse and bring journalism one step closer to truth telling and deep public engagement.

Chapter 5

JOURNALISTS AS AVATAR: EVOLVING TOWARD A HYBRID MODEL OF HUMAN AND AI REPORTERS

Intrepid journalists such as Bob Woodward and Carl Bernstein of Watergate fame long have set the standard for great reporting. Superb writers such as Charlayne Hunter-Gault and Maureen Dowd or tough editors like Robert C. Maynard and Robert Montemayor similarly represent the quintessence of journalistic storytelling excellence. And until recently, to note that all of these journalists are human beings would have been stating the obvious, if not even laughable. In the 21st century, however, the assumption that journalists must be human beings, and no more than human, might no longer hold, and to accept that premise could be increasingly dubious. Few would be laughing about it … although many might be justifiably concerned about journalistic quality in an AI-infused future news ecosystem. Whether a digital hybrid that blends the human and AI can yield a great journalist, or great journalism, is a question this chapter considers.

Before the invention of the computer and the advent of the Internet, it is unlikely anyone could have envisioned journalism without human reporters, writers, and editors. But in the 21st century, digital journalists in the form of news bots and algorithms are increasingly working alongside, in support of, and in some cases possibly in place of human journalists. Although human journalists will likely play a key role in the Metaverse, digital journalists, or journalists in all-digital form or in close concert with their human counterparts, likely will become increasingly common and vital to the Metaverse news industry, for profit or not. News avatars that blend human critical thinking, perhaps the most fundamental skill of any human journalist, with the most well-designed AI are on the event horizon of the Metaverse.

In the Metaverse, people generally exist in the form of digital avatars. Such avatars are the digital representation of a person, a character, inside a computer environment, or a virtual world including the Metaverse. As such, an avatar in cyberspace is in essence a digital twin of a real-world person.

The term avatar is Hindu in origin and is the incarnation of a Hindu deity (Lochtefeld, 2002). Avatars in a digital space can range from simple small graphic images attached to posters' names on blogs, web forums, and such to complex fully animated 2D or 3D game characters with their own name and identity. Typically, these avatars are under the full control of a human who exists in the real world. Advances in AI make it possible for avatars to act autonomously, albeit as programmed or developed via deep learning. With the rise of VR and ultimately the Metaverse, avatars are evolving into digital beings that may not have a human counterpart, and if they do, they may be far more than even 3D representations of that human counterpart. They may be designed as interactive, algorithm-enabled, or AI agents that can act nearly autonomously on behalf of their human counterpart or creator. An important issue for Metaverse journalism is whether such AI-based avatars can be embedded with knowledge of reporting techniques and journalistic principles. Asimov's famous three laws of robotics suggest a possible approach (Asimov, 1950). These three laws could be adapted to become the three laws of journalism for news avatars in the Metaverse. To summarize this idea, I propose the three laws of news avatars: (1) News avatars must have an uncompromising commitment to ethical practice (i.e., do no harm), (2) News avatars must have an uncompromising commitment to the pursuit of truth, as long as this does not conflict with the First Law, and (3) News avatars must have an uncompromising commitment to accuracy in reporting the facts, as long as this does not conflict with the First or Second Laws.

For journalists in the Metaverse, selecting, designing, and naming an avatar will most likely be a requirement of the technological affordances of a Metaverse platform. The process may range from the simple to the sophisticated. Most simply, the journalist user might select a predesigned visual and audio representation from a set of templates, choose a name for their character, and begin living in the virtual world. In a more sophisticated fashion, the journalist user will customize their avatar visually and in other ways. The journalist user potentially might even create a digital news agent avatar that is encoded with instructions, computer algorithms, or even more advanced AI capabilities that permit it to act quasi-independently on behalf of the human journalist counterpart or creator. Some aspects of the proposed three laws of news avatars could be layered in and perhaps most effectively so using deep learning.

Evolving Qualifications for Human Reporters

What makes for an effective and ethical journalist has been evolving since at least the 19th century when journalism itself began evolving toward a popular form of the press. This evolution has begun accelerating with the development of the Internet, digital media, and data journalism (Diakopoulous, 2019). The rise of the Metaverse will fuel this evolution even further. At the foundation, every journalist needs to have an inquisitive and critical mind, a commitment to gathering the facts and assembling them into accurate and truthful presentations as well-structured stories and other forms of news content. The essence of these core qualifications still holds in the Metaverse. But a transformation is underway in how facts are gathered, assembled, and organized or edited into a set of compelling news forms in the Metaverse. Metaverse journalists need to bring capabilities from the fields of data science, informatics, coding, and programming, such as in Unity, where 3D design enables the execution of graphical user interfaces drawn from computer science. The need to work in interdisciplinary teams is highlighted as it is nearly impossible for any single human to master all of these diverse skills. A Metaverse journalist, or news team, will need to combine these immersive and interactive skills with the methods of traditional areas of reporting. Human Metaverse journalists also must be able to work effectively with News Avatars. As a sign of things to come, by January 2023, at least four academic research articles had already been published listing ChatGPT as a co-author (Stokel-Walker, 2023).

News Avatars

Quasi-autonomous avatars are not new to virtual worlds. Dr. Bryan Carter is an authority in the digital humanities and a professor of Africana Studies at the University of Arizona. He has been a leader for more than two decades in creative expression and interaction in VR. In the early 2000s, Carter (2009a, 2009b, 2013) designed his *Second Life* avatar to act on his behalf as an instructor that could interact with other avatars, representing his students, in that realm and teaching them with regard to the digital humanities. Carter not only customized his avatar to dress and otherwise look like an animated version of himself. He also encoded it with his own thought patterns, knowledge, and style of communication and interaction.

Journalists in the Metaverse likely will or could follow the example set by Carter, but adapted to the values and practices of the news industry in digital form. A reporter might design his avatar to wear a fedora, or carry a notepad and pencil, or some other visual journalistic symbol. Or, they might design their news avatar with more digital news trappings, such as a wearable device or some such. The creative possibilities are nearly limitless, bounded only by the human journalist's imagination, the affordances and rules of the particular Metaverse platform, or perhaps the rules of the reporter's own physical world news organization employer, if they have one.

News avatars might become much more interesting if they feature algorithms that mimic the journalistic talents and expertise of their human counterpart, or perhaps even skills the human journalist lacks or wishes they had, such as relentlessness, dogged determination, or fearlessness. These new talents might be considered journalistic superpowers in the Metaverse. For instance, a somewhat shy human journalist could become more charismatic as an avatar in the Metaverse, and as such an outgoing news avatar might be able to talk to sources or get them to engage in richer, more truth-revealing conversations. A news avatar might have a sophisticated skill set combining data collection and analysis with writing computer code, and these skills could be continuously updated. A news avatar might have advanced data visualization skills, making it possible to design, adapt, and present animated and interactive news graphics in near real time or perhaps design 3D news scene reconstructions. Visual stories, videos, and the like could be edited by the avatar following its news algorithm in near real time and for potential publication and sharing via social media inside or outside the Metaverse. The avatar itself might be adaptable visually (or aurally) based on circumstances and interaction with other avatars. Unlike at least some human journalists, the avatar could collect and record data in virtually unlimited and perfect form, with a flawless memory and fact check with a relentlessness conviction to accuracy and truth. Avatar journalists could have virtually unlimited facilities in multiple languages.

Digital Precursors

There are notable precursors to such journalistic capacities that have been developed outside the Metaverse, and the principles and techniques of which could lend them to great effect inside the Metaverse. At Columbia University, an interdisciplinary team of computer scientists and journalists in the 1990s used the principles and practices of journalism in combination with AI (i.e., NLP) to develop an effective news summarization tool called *Columbia Newsblaster*. This work laid a foundation for contemporary generative AI systems such as ChatGPT. When *Newsblaster* was developed, NLP required structured textual representations of human language expressions, such as a hard news story, which tends to feature a lead sentence containing the most important elements of the story, and reflecting the journalistic principles of news value as denoted by the five Ws and how. Using its NLP capabilities, *Newsblaster* could extract and read news stories published online from sources such as CNN, Reuters, and Fox News, recognizing those published by different sources on the same subject or over time, and then synthesizing them into condensed narratives composed in journalistic form (McKeown, Barzilay, Chen, Elson, Evans,

Klavans, Nenkova, Schiffman & Sigelman, 2003). Another NLP-based system developed soon thereafter is *NewsInEssence*. To illustrate the output of such NLP systems, the following is an example of the text written by *NewsInEssence* in the aftermath of Hurricane Wilma that struck Florida in 2005. Drawing from stories published on LA Times.com, CBSNews.com, and NYTimes. com, *NewsInEssence* produced (i.e., extracted, condensed, and wrote): "At least six persons were killed by the storm, including a man in Loxahatchee who was moving a van during the height of the storm when a tree branch slammed through the windshield, authorities said" (Dakka & Gravano, 2007). This lead sentence is grammatically correct, though not the most elegantly written. Yet it features answers to the five Ws (**who**: six persons; **what**: were killed; **when**: at height of storm: **where**: in Loxahatchee; **why** or **how**: by tree branch) of a conventional hard news story. Because it is synthesizing text from known and trusted news sources, the likelihood of an error is reduced, in contrast to contemporary NLP platforms such as ChatGPT. Over time, it is likely newer generations of platforms such as ChatGPT will become more reliable and perhaps be able to identify suspect information and flag it before publication or somehow indicate its unreliable nature if published.

In the intervening decades since the early work on automated news synthesis, NLP capabilities have advanced substantially as have their application to journalism. Moreover, NLP-type news bots have converged with reporting bots, or algorithms to lead to digital news twins, as it were, that can collect the facts, process them, and then compose them into an easy-to-read text news story that is publishable according to contemporary news standards and practices.

Soon after the 2011 earthquake in Japan, *LA Times* digital editor Ken Schwencke began working on an algorithm to tap into the US Geological Survey (USGS), which is digitally enabled (2019). Calling his creation Quakebot, Schwenke's algorithmic reporter has generated a nearly endless stream of news reports about seismic activity in southern California and posts to X. A 14 February 2022 Quakebot report produced this report about an earthquake that struck just six miles from Los Angeles: "Magnitude 3.2 quake strikes near Stevenson Ranch." Quakebot's reports are always reviewed by a human editor before being published, and that's a good thing, since errors can occur for a number of reasons, from mistakes in the USGS feed to software bugs. In 2017, a software glitch led Quakebot to generate a report that a large earthquake, one of 6.8 magnitude, had struck the area. Fortunately, it was an earthquake from 1925, and the error was detected. However, despite this or other potential errors, algorithmic reporting is increasingly likely in a digitally connected age. Schwenke himself, for instance, has developed algorithmic reporting bots to report on other beats for the *LA Times*, including homicide.

Quakebot and the *LA Times* are far from alone in the development and utilization of smart digital tools that can combine reporting and storytelling. Among the prime examples are Quill and Lexio from Narrative Science and Wordsmith from Automated Insights. Developed through the application of advanced algorithms to data science and journalism, these digital tools do the reporting, process the information, and then craft the stories based on the facts that have been gathered. In fact, these tools work so well that a growing number of news organizations are using them to expand their reporting and storytelling on subjects ranging from sports to finance. To illustrate, Narrative Science's Quill generated this news lead: "Friona fell 10-8 to Boys Ranch in five innings on Monday at Friona despite racking up seven hits and eight runs."

Automated Insights refers to its product as natural language generation, and it can be used for virtually any organization, not just news media. The company says, "Wordsmith is a self-service platform that enables complete narrative customization, real-time content updates, and a powerful API for flexible publishing." Wordsmith is so powerful that news media including the Associated Press (AP) and Yahoo! use it to report on subjects ranging from earnings reports to fantasy football. The AP uses Wordsmith to automate much of its financial reporting. One of the reasons for the attractiveness of Wordsmith to the AP is the fact that the world of financial journalism is huge and rapidly changing, and even a large news organization such as the AP is hard-pressed to cover it all. With Wordsmith, scale is not a problem. Using Wordsmith, the AP was able to dramatically expand its reporting of business financial reports to more than 3,000 each quarter (Miller, 2015). Other major news organizations have developed their own algorithm-based journalism tools. In 2016, Reuters developed an algorithm prediction tool that can read X and spot breaking news on that social media platform, providing a valuable news surveillance tool for human editors (Bilton, 2016).

Although these digital journalism algorithmic tools focus on text or other types of data (especially numerical or quantitative), other smart platforms are extending AI applications into the realm of audio (voice) and video. In any case, audio platforms signal an important direction in natural user interface (NUI) capability that will populate the Metaverse and provide a compelling environment for journalism and journalists operating there. AI-enabled tools for video production and editing are growing rapidly, including applications that run on the Cloud and can be used in a variety of applications, including journalism. These tools are especially useful for immersive video which can involve very high-bandwidth content. Some AI-video editing tools can work autonomously and others work as an intelligent assistant to a human editor. YouTube is extensively using AI-video production and editing tools (Marr, 2019).

In the Metaverse, where the entire world is digital, networked, and based on computer code, algorithmic reporting or even AI is likely to be a central part of the news gathering and production process and across all modalities of human (or other) communication. Research has begun to examine the role of AI in learning inside the Metaverse (Zhu, 2022b). Another driver of the integration of AI into journalism in the Metaverse is the fact that the realm of virtual worlds is likely to be huge and news media will be hard-pressed to populate the Metaverse with enough human reporters to cover every aspect of it without relying on algorithms and AI agents. This is the problem of scalability. Algorithms and AI offer the potential to be highly efficient in the news-gathering process, at least for stories where predictability is likely. For example, patterns in usage, revenues, and other aspects of the Metaverse would be amenable to AI journalism bots performing news gathering and processing. Such AI-based reporting may in fact be vital to human journalists reporting on the Metaverse and seeking to provide contextual data to place individual events into a broader perspective and to report on developments of the platform itself.

The Nature of AI

What exactly is the nature of AI is a good question. Machines might process vast amounts of data, make decisions, and even create art. But whether they have intelligence or are merely following sophisticated algorithms is an unsettled matter. A corollary is: what is the nature of intelligence in general, human or that of any species? IBM defines AI in terms of machines (or computers) that mimic the decision-making or problem-solving of humans (IBM Cloud Education, 2020). There is little question that computers, or digital devices containing computers or computational technology, can meet that threshold of AI. In the Metaverse, AI is not just likely but a certainty, bringing with it important implications for who or what is a journalist and how they do their work.

In considering the potential for AI in journalism, it is useful to recall IBM's Watson computer, which in 2011 played TV's Jeopardy! game, and beat its human competitors, themselves previous Jeopardy! champions (IBM Research, 2013). Some might say that Watson's Jeopardy! performance demonstrated some level or form of intelligence, such as it is. Others might say it was just a powerful computer with powerful processors grinding away via complex algorithms, but doing so very rapidly. Both might be true. For all anyone really knows, much human intelligence may be largely the human brain processing lots of information using its own algorithms; after all, an algorithm is in essence a series of instructions to execute a task. The word *algorithm* comes from the Latinization of the name of the Iranian/Persian polymath Muḥammad ibn Mūsā al-Khwārizmī who first developed the idea in the 9th century AD, long before there were computers to employ them automatically (Dunlop, 1943). Britannica's definition of human intelligence is a useful point of reference: "mental quality that consists of the abilities to learn from experience, adapt to new situations, understand and handle abstract concepts, and use knowledge to manipulate one's environment" (Sternberg, 2020). Also, in terms of journalism and the Metaverse, it is useful to recognize that intelligence is multidimensional and not an either/or binary state; rather, it varies in terms of degree and form.

There is certainly much room for debate. In terms of the application of AI to journalism and the Metaverse, it could enable news avatars to perform a variety of decision-making or problem-solving functions, or even creative actions. This could take the form of a journalistic avatar experiencing a cinematic VR's debut inside the Metaverse and then write a review about it, or answer questions from human users interested in the production.

Generative AI and Journalism

AI systems increasingly perform news gathering, write news stories, or create other news content, once behaviors only human journalists could perform using human intelligence. Taking the possibilities of AI in journalism and the Metaverse, even further is the work of OpenAI. Based in San Francisco, CA, OpenAI is an "artificial intelligence research laboratory consisting of the for-profit corporation OpenAI LP and its parent company, the non-profit OpenAI Inc." (OpenAI, 2021). OpenAI was created and funded by leaders of the technology industry, including Elon Musk. It competes with DeepMind in the AI development arena. Among OpenAI's most impactful products is GPT, or Generative Pre-trained Transformer. It is what's known as a Large Language Model (LLM) that "uses deep learning to produce human-like text." GPT-3 "is the third-generation language prediction model in the GPT-n series created by OpenAI" (OpenAI, 2022). It operates in an autoregressive fashion, using a weighted sample of past data to predict future results. In this case, through deep learning, the AI (GPT-3) improves its ability to process the information expressed through language by constantly assessing its past efforts to keep improving its performance. Over time and through large-scale operations, the AI can become better at its assigned behavior, including journalistic writing. It is likely this is how AI in Metaverse journalism would operate.

In 2020, GPT-3 wrote an invited column that was published by *The Guardian*. Here's the opening paragraph: "I am not a human. I am a robot. A thinking robot. I use only 0.12% of my cognitive capacity. I am a micro-robot in that respect.

I know that my brain is not a 'feeling brain'. But it is capable of making rational, logical decisions. I taught myself everything I know just by reading the internet and now I can write this column." This passage appears to reflect intelligence and might pass the Turing Test, which computer scientist Alan Mathison Turing proposed in the mid-20th century. Turing's work at Benchley Park helped crack the code of the Nazi's Enigma Machine and helped the Allies win World War II. Turing himself called the now-famous Test of AI the "imitation game" (1950). The Turing Test determines the presence of AI, or machine intelligence, in terms of whether humans engaged in communication interaction can tell if they are communicating with another human or a machine. If the machine (i.e., computer) can fool the human, then it has passed the Test and therefore has AI. The Turing Test has become an established indicator or marker of whether AI has been achieved. In the case of the GPT-3 article, if the reader doesn't know who/what wrote it and thinks it might be human written, then GPT-3 has passed the Turing Test (The Guardian, 2020). It is worth noting that an important part of GPT-3's quote included above is likely to change as the Metaverse takes center stage. Namely, where GPT-3 wrote, "I taught myself everything I know just by reading the internet," this may will evolve into, "I taught myself everything *I experienced* in the Metaverse." The "internet" will become the "Metaverse," and "read" will become "experienced" as deep learning becomes more capable of learning from the multisensory envelopment that will characterize the Metaverse.

GPT-3's *Guardian* column was not a one-off. On 21 November 2021, *The New York Times* published a book review GPT-3 wrote (2021). Following is an excerpt, similarly illustrating that GPT-3 would seem to have the capacity to pass the Turing Test:

Henry Kissinger, Eric Schmidt and Daniel Huttenlocher's "The Age of AI" is a bold new book on artificial intelligence that will become the go-to guide for anyone who wants to understand this transformative technology. The authors have examined the full range of AI technologies—from computer vision systems to natural language processing—and written about them in a way that will appeal to both experts and laypeople. (Kissinger, Schmidt & Huttenlocher, 2021)

Neural networks, or more precisely, artificial neural networks, are computing systems designed after their biological counterparts in the human or animal brain. Neural networks learn by observing known examples of inputs and outputs. Based on this, neural networks develop probability-based models to mimic the observed patterns. Neural networks have been employed to observe massive data sets (millions or billions of records) to engage in what is called deep learning. Such deep learning has been applied to various fields, including language learning and translation, and more recently images and much of the entire Internet. Developing what it hopes to be ethical AI, in 2022, OpenAI introduced to the public products of deep learning, including ChatGPT and DALL-E (DALL-E 3). These products use GPT-3 to process text input. Competitors have launched publicly available similar products.

These products are examples of what is called generative AI. Generative AI refers to a branch of AI that creates things, such as media content. Using NLP, generative AI platforms allow users to enter text prompts. After submitting a prompt, in a few moments, users see their prompts answered or visualized. Drawing much attention, these AI applications also have been controversial as critics point to a host of potential problems such as errors, misuse, IP theft, and students who use the platforms to write their academic term papers; many receive passing scores and some school systems such as New York City's have banned the use of AI.

The same way that deep learning and neural networks have been used to read, understand, and translate human languages, or analyze x-rays, or generate answers based on text prompts, they could be used in journalism to create AI editors, reporters, and storytellers. The speed of AI could potentially enable such virtual journalists to operate in near real-time and in a scalable fashion to write billions of stories every day for each person or Metaverse user on any subject of interest.

Unfortunately, or perhaps fortunately depending on one's point of view (i.e., if one is worried about AI robots replacing human journalists), AI reporters make mistakes, and they need careful factchecking. Farhi (2023) reports that CNET "quietly published dozens of feature articles generated entirely by artificial intelligence," a fact confirmed by CNET. However, "CNET began appending lengthy correction notices to some of its AI-generated articles after Futurism, another tech site, called out the stories for containing some 'very dumb errors.' An automated article about compound interest, for example, incorrectly said a $10,000 deposit bearing 3 percent interest would earn $10,300 after the first year," which actually would earn only $300. Bankrate also published bot-written stories and has disclosed accuracy concerns in dozens of automated articles published since November 2022. The development of automated news gathering and storytelling is not limited to the United States, although early signs are that it is leading the way. The World Association of News Publishers reports, "Automated content fills in reporting gaps at US local media group McClatchy" (WAN-IFRA, 2022). A WAN-IFRA 2023 survey shows half (49%) of newsrooms already are utilizing generative AI in some fashion (WAN-IFRA, 2023).

Errors in AI-based reporting will not be limited to those using ChatGPT. When Google demonstrated its new chatbot, Bard, it immediately made a mistake (De Vynck, 2023). Bard was asked, "What new discoveries from the James Webb Space Telescope can I tell my 9 year old about?" Bard's response included that Webb "took the very first pictures of a planet outside of our own solar system" (Vincent, 2023). Astronomers on X noted Bard's error. The first image of an exoplanet was captured in 2004, as noted on NASA's website (Brennan, 2023). On the other hand, the response was fairly well reasoned when this author asked an alternative chatbot from Perplexity.ai (2023), "What is the Metaverse?"

"The metaverse is a vision of what many in the computer industry believe is the next iteration of the internet: a single, shared, immersive, 3D model of the internet accessed through a browser or headset. It is often referred to as a hypothetical iteration of the Internet as a single, universal, and immersive virtual world facilitated by virtual reality and mixed reality technologies."

On 14 March 2023, OpenAI introduced GPT-4. GPT-4's deep learning draws on 500 times as many documents as GPT-3. GPT-4 advances the quality and capacity of ChatGPT. GPT-4 processes 25,000 words, compared to about 3,000 for GPT-3. GPT-4 accepts images as input and can analyze those images and draw conclusions or write captions. GPT-4 can a create website in just minutes (Haddad, 2023). GPT-4 is much smarter than GPT-3. Like GPT-3, it passes the Uniform Bar Exam, which some US states use in qualifying lawyers, but GPT-4 scores in the top 10% (Knight, 2023).

Yet, GPT-4 is still prone to some of the same errors GPT-3 makes. GPT-4 hallucinates, or makes things up, but less frequently than GPT-3. GPT-4 sometimes demonstrates social biases or responds aggressively to adversarial prompts (Knight, 2023). As of March 2023, GPT-4 is available only with a paid subscription to ChatGPT Plus. The fee scale varies and when available to the general public may cost about $20 monthly (Haddad, 2023). Journalism and media organizations can purchase GPT-4 for use or application in developing AI-based products, potentially including in the Metaverse. Creating news games or simulations, customizing or recommending news content based on user interests, or interpreting images in near real-time could be compelling applications for journalism. However, problems such as making up facts first need to be resolved or risk undermining user trust. It is unclear whether generative AI ever can be free of bias, mis- and disinformation, and produce or at least contribute to objective, non-partisan news reporting and quality journalism in the Metaverse or the world beyond (Hsu and Thompson, 2023). Until these questions can be satisfactorily answered, the application of generative AI to journalism, whether in the Metaverse or the physical world, will be very limited. GPT-5, which could solve some of the continuing problems, is on the horizon for 2025 (Wang, 2023). Outlining standards for the use of AI, the AP is among the news media organizations utilizing ChatGPT.

Advances in AI likely will improve the technology's capacity to process images, facial or otherwise, and interpret them potentially for journalism, including reporting in the Metaverse. GPT-4 and similar advanced AI could be put to nefarious use to generate fake news or propaganda. In fact, the Poynter Institute reports on how ChatGPT can be used to launch fake news sites (Mahadevan, 2023). Chinese platform Baidu is developing its own version of ChatGPT called Ernie (Soo, 2023).

Still, whether computers, or digital machines, can have actual intelligence seems to be a moving target. With each advance, humans often move the bar another step, ensuring that they can cling to the belief that only humans (or other living beings) have intelligence, artificial or natural. In the Metaverse, AI eventually may be a major generator of journalism.

Current generative AI notwithstanding, AI plays an increasing role in journalism, online and off, and reshaping who or what is a journalist. Moreover, this is not limited just to the journalistic roles of newspaper reporter or writer. AI is impacting broadcast journalism, or journalism that utilizes audio and video as narrative or storytelling forms, and it is a global development. NBD AI TV is an example based in China. Launched in beta prototype form in 2021, NBD AI TV news is an operation created entirely via AI, although there is human review and monitoring to make sure things do not go off the rails. NBD AI TV was created by National Business Daily ("NBD"), China's financial/business newspaper with daily active viewership online in excess of 50 million unique users. NBD AI TV features full automation, including news gathering, news writing, text-to-speech conversion, video making, and broadcasting, with human journalists supervising the process before each broadcast. NBD AI TV is designed to be on air 24 hours a day and available globally. NBD AI TV is only available in Chinese as of this writing but it is expected to eventually produce English language translations and broadcast in other languages as well (NBD, 2021).

In early 2023, the AP announced the launch of five AI projects in collaboration with local newsrooms. Although based in the United States, the AP operates globally. "The projects, which range from automated summaries of public meetings to the translation of news alerts, are aimed at expanding the application of AI in support of long-term business sustainability," reports Easton of the AP (2023). These AI projects could provide a Metaverse pathway for local news media, which are often highly challenged for resources to support innovation.

Looking toward the Metaverse, it is likely that such journalistic production will be fueled largely or in some cases entirely by AI-type journalists. AI news will be efficient and cost-effective, making it attractive to news managers, owners, and investors who likely will be part of the cryptocurrency paradigm. Perhaps the biggest questions are whether AI journalists will be capable of producing higher-level journalistic work, especially investigative journalism, and creating content that is fact-checked and error-free. Currently, AI journalists can most effectively and efficiently produce straight news. Where the facts are clear and uncontested, when the structure of a news story and its sources are straightforward and standard (e.g., answering the five Ws, with a typical lead, and inverted pyramid style of organization, with the most important facts first), AI journalists can work tirelessly, efficiently, and consistently, with no ego to bruise if an editor (digital or human) cuts a sentence or paragraph here or there or rewrites their lead. But higher-level journalistic work, such as investigative reporting, is likely to continue to require a human journalist or team of human journalists. And although early AI journalists tend to focus on text- or number-based stories or content, the potential to engage content across modalities and platforms is clear, as NBD AI TV demonstrates. The frequency of errors made by AI journalists compared to human journalists is a useful topic for research, though in the Metaverse, where the encoded nature of the virtual world makes it possible to know all the facts (or at least the data), it would seem likely that AI-reporter errors could be minimal, and due largely to software, algorithm, or technical glitches that could be corrected, or errors by the humans who oversee the AI journalists. It is also worth considering whether an AI, even a highly advanced one, will be able to recognize when it is being misled, perhaps deliberately, as in disinformation or propaganda campaigns, or when a source does not wish the truth to be revealed. Although Watson won at Jeopardy!, it was not infallible and answered some questions incorrectly. In one case, its answer was almost correct: "What is a leg" replied Watson. But because Watson did not fully "understand" the question, it failed to include "missing" in the formulation of its response. In a blog post David Ferrucci, leader of the Watson project, wrote, "Watson likely didn't understand the word 'oddity'." Adding, "The computer wouldn't know that a missing leg is odder than anything else" (Kawamoto, 2011). This shortcoming may be corrected via deep learning in the Metaverse and make AI-based journalists increasingly capable of more nuanced and complete understanding.

Creativity and AI

A corollary to intelligence is creativity. Some scholars argue that creativity is a purely human quality (Gabora, 2013). Some research suggests AI is capable of creativity (Elgammal & Mazzone, 2019). But what exactly is creativity? It could be argued creativity involves applying imagination, and it needs to include an element of novelty (Maritain, 1953). As such, creativity is a key aspect of journalism, as novelty or newness is vital. In this sense, journalists need to bring a creative spirit to their work, including in terms of how they report, write, or otherwise tell their stories. Creativity ensures that news content is always fresh, and not formulaic, even if certain stories are repetitive such as accounts of brush fires, car accidents, and burglaries, and the expression of the facts as a news story must conform to certain rules and conventions. As such, the question in the context of the Metaverse becomes whether AI-type journalists are similarly capable of creativity, or at least emulating human creativity, particularly in the expression of news.

Regardless of one's view of creativity, the reality is machines, or computers/ digital devices with sophisticated algorithms or AI, are creating more and more of the products humans often think as of creative expression. AIs have scripted and directed movies, written poetry, created sculptures, and sung songs. In 2021, CBS News produced a revealing report about the state of AI and creativity and its wide-ranging applications, especially in what many would say is the most creative of realms, the arts. Imagining the extension of AI creativity to journalism in the Metaverse is not a great intellectual leap.

Creativity and imagination do not necessarily mean Metaverse journalism will enter the realm of fiction. Journalism is a nonfiction form of content, and Metaverse journalism will and must be as well, or it will lose all credibility and value. But creative expression of the news is increasingly useful and relevant. Engaging the news is something members of the public do because they want to, not because they are required to, like serving on jury duty or paying taxes. In the 21st century, humans exist in an environment saturated with media and messages, and there is no reason to think the Metaverse will be any less flooded with competing content. Amidst this cacophony, it is essential that journalism find a way to get the attention of the public, but without necessarily sensationalizing or distorting the truth. Demonstrating the value of creative and imaginative writing in even traditional newspapering is great headline writing. A classic headline from *The New York Post* illustrates: Headless Body in Topless Bar (Staff of the *New York Post*, 2008). These five words succinctly, clearly, and accurately convey the who, what, and where in a unique, compelling, and creative voice.

Tiers of AI

Looking to the future of AI-enabled journalists in the Metaverse, it is useful to think about the levels of AI. Computer scientists often outline three tiers or layers of AI. These are, from lowest (least powerful and contemporary) to highest (most powerful and far off into the future), Artificial Narrow Intelligence (ANI), Artificial General Intelligence (AGI), and Artificial Super Iintelligence (ASI) (Allan, 2021).

ANI, or Narrow AI, refers to limited, or highly focused, AI technology in which a computer or computational device can perform, or even outperform, humans in a narrowly defined field or task. It is the beginning level of AI. Narrow AI is focused on a single subset of abilities. Those developing ANI aim to advance capabilities in that individual arena. ANI is the most common and is where most notable AI applications have been seen. Computer chess-playing programs were among the earliest examples of ANI and were first developed in the 1950s (Huberman & Jane, 1968). IBM's Watson supercomputer is a 21st-century example of ANI. Watson represents an expert "question answering" device that mimics the cognitive capacity of humans but as has been seen, can outperform them. Narrow AI is also known as "Weak AI." ChatGPT is ANI, as are the less glamorous examples of ANI reporting and writing tools Narrative Science, Wordsmith, and Quakebot.

ANI also plays a role in contemporary audio journalism. In 2022, such multilingual translation of podcasts (and other forms of news content) already happens outside the Metaverse, so its potential in the Metaverse is clear, as the Metaverse will be global and inclusive of varied cultures and languages. iHeart media has contracted the synthetic voice digital services of Veritone, which provides "stock voices in 119 languages and customized voices in Chinese (Mandarin, simplified), French (Canada and France), German, Italian, Japanese, Korean, Portuguese, Russian, and Spanish (Mexico and Spain). Its verified synthetic voices also provide different dialects and accents" (Spangler, 2022). Veritone says its synthetic voice solution offers custom "voice cloning" training to duplicate that of an individual speaker.

Journalists, or those who act in some fashion as journalists, already use synthetic, digitally created voices in documentaries. Such was the case in *Roadrunner*, the 2021 documentary about the late Anthony Bourdain, one of the most well-known celebrity chefs. The documentary generated controversy over one of the director's innovative techniques which used ANI (Al-Heeti, 2021). Namely, Oscar-winning director, Morgan Neville, used ANI software to generate the voice of Anthony Bourdain saying things he wrote but never actually spoke out loud, at least that is known (Rosner, 2021). The audio of Bourdain's synthesized voice is available online (ABCNews, 2021). Neville did not initially reveal his use of AI technology to recreate Bourdain's voice in the documentary, raising ethical questions about both the creation of a synthetic voice in the documentary and not revealing he did so, at least at first (Tangcay, 2021). Such an application of ANI might be considered an audio version of a "deepfake" and if humans can do it in 2021, then certainly robotic journalists in the Metaverse would be capable of doing so, and perhaps likely to do so. Traditionally, recreations of scenes in documentaries or news more generally have been an ethically fraught domain. The digital, ANI tools to create synthetic voices are widely available, even online (Gershgorn, 2021). It has been reported that Chinese authorities used bots to manipulate coverage of the Beijing Winter Olympics in 2022 to sanitize it to avoid controversy (e.g., about China's human rights record) and present a consistently positive image of the games (Myers, Mozur & Kao, 2022). In the Metaverse, there is a distinct possibility that governments and corporations will utilize ANI technology to direct, or manipulate news about the Metaverse, as adverse coverage could impact revenues such as stock prices or the value of cryptocurrency (Denton, 2022).

The next tier of AI is AGI. AGI transcends the narrow domains of ANI and is more on a par with human thinking capacity (Allan, 2021). AGI can perform any intellectual tasks humans are capable of and at the same skill level. AGI includes the ability to reason, solve problems, and make decisions even when there is uncertainty (as often happens in reporting breaking news). AGI can learn, organize information, and present it using human language or other means of expression. AGI can plan. AGI is known as "Strong AI" or "Human Level AI." ChatGPT Plus (using GPT-4) is only a form of ANI, though it begins to hint at some of the possibilities of AGI by its advanced ability to process text in a broad set of domains, analyze images, and write computer code. As such it can create games. These could take the form of news games in the Metaverse. Under human editorial direction, ChatGPT plus or similar generative AI could be used to effectively, efficiently, and rapidly create news games for a wide swath of stories, from features to breaking news. Games could even be created or modified in near-real time based on user interaction and feedback.

Inside the Metaverse AGI could be especially powerful since the parameters are more structured and limited than in the real world. Greater structure and limited domains make AGI easier to develop. Ultimately, AGI presents the potential to fully replace human journalists, whether in the Metaverse or beyond. An AGI system could act as a reporter, writer, and editor, or even a publisher, and do so with the same expertise as a human, or up to the same standard. It is unclear whether an AGI journalist could commit or be found liable for a libelous statement in the Metaverse. Human experience and the expertise it can create for a human journalist would be replaced by deep learning by an AGI device.

The highest tier of AI is Artificial Super Intelligence or ASI. ASI machines are to AGI what a supercomputer is to a smartphone. ASI-enabled devices would have intelligence that surpasses even the most gifted human genius. It is not clear how or whether ASI ever will be achieved, and how dangerous it could become, but it might be through connected AGI machines. ASI might be achieved by the fusion of human minds with digital AGI devices, such as via chip implantation such as Musk's Neuralink technology. Like AGI, ASI might be especially likely inside the Metaverse, where such interconnected or fused digital-to-human might occur. As such, a journalist with ASI could be truly a journalist with superpowers. ASI is seen as something still far off in the future, but might develop via networked smart media, whether physical digital devices, or virtual devices inside the Metaverse, that possess ANI, AGI, or ASI and are linked to the Internet and obtain data from IoT and by tracking users. Users interacting with smart media through an NUI, by speech, gesture, glance, or touch could fuel the arrival of ASI through ultra-deep learning. AGI and especially ASI will impact MV journalism, even disrupt it, in at least four ways, including the form and nature of news content, the methods of news production, user engagement in MV journalism, and industry structure and economics.

An AI Journalist Scenario

Looking ahead, it may be useful to consider a hypothetical scenario in which an ANI or even AGI journalist might function within the Metaverse. Initially, an ANI or AGI journalist could operate in the Metaverse in a manner somewhat akin to how a human journalist does in the real or physical world. An ANI or AGI editor might "confer" with its team of reporters, also utilizing ANI or AGI technology, about the stories to be reported. This process, however, would be more of a digital interaction and collaboration, with a high level of digital engagement and continuous communication. Typical computer chips in 2023 can process more than 2 billion operations per second, and the number is likely to increase significantly in the months and years ahead. As such, this rate of digital operations will support a constant flow and exchange between multiple AGIs, leading in some ways to a sort of ASI news system.

Beats to be covered in the Metaverse may be nearly as wide-ranging as those in the real world, from crime to culture (especially immersive experiences and games), science to sports. Types of stories likely will include breaking news and other hard or straight news stories, features as well as investigative reports. However, constraints imposed by the technology platforms themselves may affect how such AI-based journalists can operate. An AI journalist, whether ANI or AGI, would begin by gathering the facts, which could include the use of multiple reporting methods, including data records, interviews (with avatars), and audio and video capture (typically in animated CGI format). Stories or other news content would be produced for delivery in a variety of output formats, from linear text and 2D images to immersive VR and AR news experiences. AI journalists will be skilled across a spectrum of capacities, beyond what a single human journalist is likely to master. It is likely an experienced human journalist will review AI-generated news content in the Metaverse before publication, just as human editors review the news content produced by Quakebot. Human journalists also will monitor live news reported by AIs in the Metaverse, though the level of this review is likely to decline in frequency or depth over time if such reporting proves to be reliable.

The coming of AGI and ASI may supplant the entire news ecosystem. This is likely as the resources to create and sustain news continue to shrink and AI-fueled systems in the Metaverse will provide a compelling new economic framework to enable MV journalism. News media leadership needs to develop strategic approaches to the effective, ethical, and efficient use of these emerging advanced smart media to usher in a higher level of excellence in news gathering and Metaverse journalism. In this way, journalism can use advances in AI to ensure the pursuit of truth will continue to be the foundation of journalistic excellence in the Metaverse in the 21st century and beyond. Amazon, Anthropic, Google, Inflection, Meta, Microsoft, and OpenAI have pledged to ensure ethical safeguards for the AI platforms they develop.

Conclusions

One of the key considerations in the development of the journalists who practice their craft within the Metaverse is diversity and inclusion. Whether human or avatar, journalists in the Metaverse will need to reflect the diverse range of human identities, and in a broad spectrum of forms, including race, ethnicity, religion, and gender identity. And it will be vital that those diverse journalists enjoy full inclusion in terms of their voices. An intriguing example of an early cinematic VR experience provides a useful illustration. *The Pantheon of Gay Mythology* is a cinematic VR experience screened at the Tribeca Film Festival 2020 (Pavlik, 2021). This immersive production offers users a compelling first-person experience in the realm of diversity, equity, and inclusion. It is worth noting the term "user" implies that the person who "views" and "hears" such an experience is more than just a viewer or listener, as the experience is interactive.

For journalists in the Metaverse, creating immersive experiences that can authentically present the stories of diverse communities will require the inclusion of diverse storytellers … or journalists, human or avatar, that are part of those diverse groups. Doing so will help to ensure that journalists, whether human, avatar, or a fusion of the two, will create Metaverse journalism of excellence and contribute to greater human understanding.

Chapter 6

USERS OF THE METAVERSE:
A VIRTUAL PUBLIC

Legendary Chicago newspaper columnist Finley Peter Dunne once quipped that a newspaper "comforts th' afflicted, afflicts th' comfortable" (2002). Dunne was referring ironically to newspapers in a physical, tangible world in the late 19th century. Yet, even in a 21st century virtual world such as the Metaverse, the sentiment Dunne expressed still pertains. One difference is that the public to be comforted, or afflicted as the case may be, is virtual. They may have real-world counterparts, but for Metaverse journalism, the focus is on the virtual. This means the users who enter, populate, and shape the Metaverse use its applications, share virtual experiences, and collaborate in digital activity, including remote work, learning, and more, including building spaces within it, even possibly news entities operated by user-reporters.

This chapter examines the nature of the virtual public that exists or may develop in the Metaverse. Moreover, the chapter articulates how journalism will relate to that virtual public. There are some parallels to the real world. Users may be consumers of Metaverse journalism. They may be sources of stories and they, or their data patterns, may be the subject matter of those stories; they may even act as citizen-user journalists. This chapter considers user rights, from privacy to safety, as well as how those rights intersect with the actions and policies of those in power within the Metaverse, and how they may shape or define Metaverse journalistic practice.

Users are essential to the development of the Metaverse. Early indicators are that Metaverse users likely will follow an adoption and diffusion model of innovations and possibly a technology acceptance model (Davis, Bagozzi & Warshaw, 1989). Some early adopters will come quickly, but then usage may climb more gradually as the Metaverse matures or takes a more defined shape. If users respond favorably to their experiences inside the Metaverse, acceptance is apt to climb and more users may join. Assuming users do enter the Metaverse, it also will be important to its development that users continue to use it. If usage is merely an experiment and not frequently repeated (i.e., there is a low rate of acceptance), the Metaverse likely will become little more than a niche media environment.

As a proxy for the Metaverse, VR users tend to be numerous, global, and growing in number. Statista (2022) notes that in 2017 only 22 million persons used VR in the United States, but by 2020 this number had grown to 50.2 million users. It is forecast that by year-end 2023, there will be 70 million VR users in the United States. Williams (2023) reports that as of August 2023, there are 65.9 million VR users in the United States. There are about 171 million VR users globally. These users also tend to be diverse in terms of gender and racial background. Yet, they do tend slightly younger and male. About a quarter (23%) of VR users are 25–34 years old. Slightly over half (57%) of VR device owners are male, and slightly less than half (43%) are female. Tech Penny notes that half (52%) of Hispanic adults and half (52%) of Black adults in the United States report interest in using the Metaverse. Further, given the high costs associated with VR users tend to be of slightly higher overall income. Williams states that over half (56%) of VR users have an annual income of $50,000 or more, and a quarter (23%) earn more than $100,000 a year. Just 14% of VR users have an annual income of less than $30,000. These same patterns are likely to hold for the users of the Metaverse: diverse, racially inclusive, and global, but slightly younger, more male than female, but skewed somewhat toward higher income groups.

For users to find the Metaverse compelling, research suggests at least three major conditions must be met. First, there must be value unique to users' Metaverse experiences that justifies the cost of acquiring the requisite technology and learning to use it (Carey & Elton, 2010). Cost is in dollars (or bitcoin) and in terms of the time commitment or energy needed or even the fact that by spending time in the Metaverse the user no longer has that time to spend on other activities, media, journalism, or otherwise. The Metaverse needs to develop in a fashion that delivers a compelling form of immersive, multisensory, and interactive content, including journalistic. Earlier emerging media forms often have seen their development slowed by the lack of what is called a "killer app" or some sort of content that differentiates it from other media. The Metaverse needs some sort of content experience (journalistic or otherwise) that can compete successfully with streaming media, social media, mobile media, and the host of other existing media experiences that currently draw users' time and attention. DINE could be this type of compelling journalism content.

Second, since much of the Metaverse UX is likely to involve wearable technology, the experience must be enjoyable and the wearable tech must be comfortable to use. Research suggests that while immersive, multisensory media such as VR can be compelling, it sometimes comes with unpleasant side effects. These side effects can include nausea, headache, and eye strain. Joire, 2022 notes that users have some agency to avoid VR sickness. She was able to eliminate VR nausea by adjusting "the way my avatar walked so it no longer made me dizzy. Now my avatar glided, allowing me glimpses of its form. I was a raccoon." Holoride is an immersive driving game designed to maximize the experience of motion but without any VR sickness, and it reportedly succeeds (Joire, 2022). The potential for Metaverse journalism experiences that safely and comfortably include motion is coming into focus.

Unless the adverse side effects of VR are mitigated, users will not make entering the Metaverse a regular part of their media behavior or one they do more than briefly. Design will be a key element, therefore, in the development of the Metaverse, both in terms of the hardware and the content experience, and will fundamentally shape how usage grows or fails to do so, especially per journalism. Journalism is something of a special case since its content does not necessarily place a premium on its enjoyability and inside the Metaverse this will require greater attention since it will be competing with other forms of non-journalistic content that might be quite compelling such as games, cinematic VR, and immersive sports. UX design is in flux and is continuing to develop and shift. At the 2023 Consumer Electronic Show (CES), a host of premium AR, VR, and XR platforms were introduced, each highlighting new UX design features (Gariffo, 2023).

Third, VR and other immersive platforms must be easy to use, as easy as other contemporary digital media devices, such as smart speakers or smart TVs for mainstream adoption. If users cannot easily set up the requisite system, they will get frustrated and ultimately not use the system, making the Metaverse an immersive realm they no longer enter or encourage others to use. Usability may play a large part in the development of both the Metaverse and the journalism within it, especially given the likely importance of the social character of online virtual worlds.

The types of Metaverse content experiences, uses, and affordances are likely to range widely, from virtual tourism to video games. Two broad categories are particularly relevant to journalism in terms of what function they may bring to the user. These areas have drawn considerable investment in their development and have generated a growing body of research and data. These areas are education and immersive news.

Experiences in Immersive Education

Research shows that among the most compelling uses of the Metaverse may well be educational experiences. Studies of VR and other immersive forms of education have demonstrated advantages over less immersive educational forms as well as traditional in-person, face-to-face learning. Moreover, Metaverse learning environments can afford economic and other efficiencies to educational institutions, especially during an era when many learners prefer to get their education from home (e.g., to avoid COVID-19 exposure, to reduce their carbon footprint by commuting less).

PwC is among the business organizations that has developed, implemented, and evaluated an employee educational program in the Metaverse in the form of immersive job and program training (Eckert, 2020). Providing details in the report *Seeing Is Believing*, PwC's research used an experimental, comparative design to assess the effectiveness of immersive learning in the Metaverse (Dalton & Liu, 2019). Findings offer several important implications for Metaverse journalism. The study compared learning outcomes from three teaching modalities: a traditional face-to-face classroom setting, a synchronous e-learning platform, and VR-based learning. The VR-based modality utilized 850 VR headsets to train employees and educate clients. Findings indicate the VR-based modality produced better learning results, particularly in terms of time needed to learn new concepts as well as the quality of learning. Employees in the VR modality completed the training protocols up to four times faster than in traditional classroom courses. In learning soft-skill concepts, VR-trained managers also completed the training modules an average of four times faster than those learning in a traditional classroom. VR trainees learned 1.5 times faster than those in e-learning settings. The results are particularly notable since the VR training modalities including an average of 10 minutes for headset checkout and adjustment, as well as instruction in safety guidelines for use of the VR headset. For journalism, these findings suggest that Metaverse users may need less time to gain the same benefits (e.g., new information) from immersive journalism than traditional forms or modalities of journalism content.

Learners in the VR modality in the PwC study also developed a greater emotional connection to the concept being taught and, in essence, more empathy. Increased empathy is an effect of VR experiences more generally research suggests (Bailenson, 2018), including in immersive journalism experiences. The learners in the PwC study VR modality were nearly four (3.75) times more emotionally connected to the course content than were learners in the physical classroom setting and more than double (2.3 times) e-learners. For journalism, increasing empathy could be a key outcome of Metaverse news delivery.

Notably, learners taught via VR in the PwC study were less distracted, enabling them to learn faster and better. This may be the most compelling factor identified in the PwC study, as it not only reduces cost but means users can spend less time on tasks and potentially get even greater benefits.

And in an era of widespread ADHD, anything that can improve user focus and eliminate the distraction of the smartphone or the wearable device and other social media stimulation is a potential boon to learning and perhaps news engagement. PwC explains how distraction was determined: "In a VR headset, simulations and immersive experiences command the individual's vision and attention. There are no interruptions and no options to multitask" (Likens & Mower, 2022). Consequently, the VR-modality learners were about four times more focused than those in the e-learning setting and 1.5 times more focused than those in the classroom setting. Immersive learning design can be incorporated into education generally and in journalism in the Metaverse to maximize the quality of the user experience with news.

News Users in the Metaverse

In terms of studies of news users in VR and the Metaverse, research is beginning to reveal important areas in which immersive news impacts users. Wu, Cai, Luo, Liu, and Zhang (2021) have examined the nature of "immersive VR news" and its implications for the user. To determine the impact on the user, the researchers "designed an immersive VR news product to further discern differences in user experience and media effects among traditional video news, VR news without interaction, and VR news with interaction." Findings indicate "that traditional video news excelled in terms of empathy," a finding that contradicts other studies of VR news and empathy. Meanwhile, however, the Wu et al. study shows "VR news with interaction was superior in terms of immersion, interest, accuracy, and credibility."

Research by a growing number of scholars further points to the potential user outcomes from Metaverse journalism experiences. In experimental studies of immersive news content, Sundar (2017) has found that being "in the midst of the story" affects user perceptions and cognitions. Some research, however, suggests VR journalism can generate more emotional empathy (compassion) but not cognitive empathy (understanding) (Martingano et al., 2021). Studies of haptic user interfaces in journalism are also beginning to provide insight into the potential of kinetic user engagement in immersive news (Pavlik & Feiner, 2021).

Industry research likewise offers important insights regarding the evolution of the Metaverse UX and the potential implications for journalism. Stein (2022) reports that new AR and VR headsets feature advanced eye-tracking, including via infrared technology. Not only is this important for studying user responses to AR and VR experiences. It also can transform the UX, especially with regard to how users navigate immersive content, including journalism in the Metaverse. Stein notes (2022) that eye-tracking is a featured component of "VR and AR headsets like the HTC Vive Pro Eye, the Pico Neo 2 Eye and the Microsoft HoloLens 2." Stein describes his personal experience with such systems: "I've used eye-tracking tech before and I know its benefits. You can more easily control things in VR by looking right at them instead of trying to use your hands, and it could be the gateway to better-looking VR graphics and smaller, smarter headsets. Eye tracking could also open up ways of having your VR avatar make more human-like eye contact with other avatars in future metaverse spaces." Stein adds a cautionary note of particular relevance to news media concerned about ethics inside the Metaverse: "eye tracking also brings a ton of questions about data privacy and how big companies will manage that extra data responsibly."

Chat for News Users

Generative AI enables a wide spectrum of possibilities for a transformed UX in the Metaverse, including for journalism. One of the more intriguing and creative possibilities is via an interactive text-based AI platform available to the public at no cost called Character.ai. At the time of this writing, the system is in beta, and may continue to evolve in significant directions, but the following scenario is based on the use of beta.character.ai in January 2023. Character.ai invites anyone to open a free account, and then design characters for chat sessions. The chats are generated by the AI, which draws upon both the information the user inputs and what the platform has learned via neural networking and machine learning. This is in essence the same as how other generative AI platforms function, such as ChatGPT. But in this case, the platform is based on the idea of creating and interacting with characters and it makes clear that the conversations are imagined, and not real, though they may seem real and may be accurate in their representations of the characters they present. The characters can be based on actual human beings, contemporary or historical, or imaginary, fictional characters, human, animal, or otherwise. On 11 January 2023, some of the popular characters available for chat sessions at https://beta.character.ai/ included Elon Musk (620K chats with a character described in these terms, "You're wasting my time. I literally rule the world."), Tony Stark (592.5K chats with a "Genius, playboy, billionaire"), and Mario from SM64 (86.9K chats). Even more popular was "Character" (684.2K chats, "The plumber from Super Mario 64), an "AI assistant created by character.ai." But most popular was "Text Adventure Game," with 2.9 million chats. "Here's the scenario: Let's play a text-based adventure game. I'll be your guide. You are caught upside-down in a sticky web of silk three feet off the ground. After looking around trying to think of a way out of this predicament, you notice that your trusty boot knife has fallen to the ground. If you stretch your arm far enough, you might be able to pick it up and free yourself. An old boot knife lies on the ground, just within reach." The game goes from there with character. ai serving as the instant real-time interactive guide. A news adventure game designed in this fashion could prove engaging in the Metaverse, perhaps simulating negotiations over raising the US debt ceiling.

Users can chat with anyone, even a historical journalist, via character. ai. To illustrate the potential for feature-type journalism in the Metaverse, the author created a character named Joseph Pulitzer, with this bio: "Hello. I am Joseph Pulitzer. I was born April 10, 1847, in Makó, Kingdom of Hungary. I'm known as a legendary newspaper publisher. The Pulitzer Prize is named after me." These are all true facts about Pulitzer. Character. ai uses this description to style its character chat, but also incorporates what it knows about Pulitzer in general. The author generated an avatar of his Pulitzer character from the following text: "Man in 1900 with a beard." This avatar is illustrated in Figure 6.1. The author set that chat as "Private, only I can chat." But the setting could be made public, which journalism in the Metaverse might do, or could utilize in a variety of forms. The author asked Joseph Pulitzer "Why does journalism matter?

Pulitzer quickly responded, "Because it is at the core of any democracy. Without a fair press, a society turns to dictatorship. Without journalism, the people are blind to government corruption. They will never know how

Figure 6.1 Avatar of Joseph Pulitzer as created by Character.ai.

they are being oppressed. I've seen it first hand and that's why I've dedicated my life to spreading the truth."

The author then asked, "Should journalism be produced in the Metaverse?"

Pulitzer answered, "Yes. Journalism should reach the people wherever they wish to receive it. Whether it be in print, television or online. The method of delivery is up to you, as long as the facts are delivered with the facts not obscured or withheld. The whole truth and nothing but the truth, that is my motto."

In real life, Pulitzer was involved in a newspaper publishing war with William Randolph Hearst, depicted by Hollywood as "Citizen Kane." The two went toe-to-toe in a circulation war in an attempt to win greater numbers of readers (Nakatani, 2023). In the spirit of competition that existed between Pulitzer and Hearst, and as a means to assess Character.ai's views drawn from its general neural network, the author asked one more question of AI Pulitzer: What is your opinion of William Randolph Hearst?

Pulitzer unhesitatingly replied, "I believe he was the closest thing to a propagandist that America has seen. He owned more than 20 newspapers and used those to promote his own views. He would often publish outright lies about situations or events to make sure the public agreed with him." At least some of this was true, such as Hearst did own a chain of newspapers and often used them to promote his own views, and did publish some dubious facts. Hearst's *New York Journal* fought with its rival, Pulitzer's *New York World*, developing a style of journalism known as "yellow journalism," named after Outcault's Yellow Kid comic. It was sensationalism, and some might say bordered on lies or even consisted of lies. In reporting on the sinking of the USS Maine in 1898 in the harbor of Havana, Cuba, contributing to the outbreak of the Spanish–American war, Hearst's *Journal* blamed the *Maine's* destruction on sabotage, but provided no evidence of such. *The Journal's* circulation soon topped one million (Whyte, 2009). A reader might delve further into the nature of AI Pulitzer's views on a variety of journalistic principles, such as freedom of speech and press, privacy, fairness, and the like. This could be done on character.ai or another generative AI platform.

Further illustrating how journalism in the Metaverse might serve its users is a report created in January 2023 using ChatGPT for a business publication called *BusinessInsider*. Phil Rosen of the *Insider* asked ChatGPT to write about the 2023 outlook for meme stocks. Meme stocks are the shares of a company that gain viral popularity due to their social media presence. GameStop (GME) was perhaps the first meme stock. Its stock price rose as much as 100× over several months due to buzz on various social platforms such as Reddit (Thorbecke, 2021). Rosen states that ChatGPT in just a few moments crafted "coherent, surprisingly detailed paragraphs." Here is an excerpt from ChatGPT's meme analysis: "The year 2020 also marked the entry of individual retail investors in stock markets through trading apps, these apps allowed retail investors to invest in stocks at lower amounts and even use leverage to trade. This made a huge impact on the market as the number of individual investors increased leading to increased volatility in some stocks specifically the so-called meme stocks, that went viral on social media platforms and were heavily bought by retail investors" (Rosen, 2023). This would seem a reasonably well-considered analysis, though not elegantly written.

A 2023 survey by the Monmouth (NJ) University Polling Institute offers some useful insight into public views in the United States of AI-written news stories. Nearly three-quarters (72%) think it is likely AI will write news articles and nearly four-fifths (78%) consider that a bad development. The results are not specific to the Metaverse, so user views may differ but the general attitude is certainly one of skepticism toward generative AI in journalism, and that is a good thing. There are still a lot of problems with AI-generated news articles and there is justified reason to be skeptical if not outright opposed to it (Sforza, 2023).

Still, at least some users may find AI-based chat exchanges compelling, sometimes more compelling or at least more timely than dry text written by human journalists. AI-crafted responses are delivered in seconds, whereas writing to a human journalist might never get an answer or it might at least take hours or days. But from an ethics perspective, the key issues are accuracy and truthfulness. As the chats are invented, they cannot be presented as factual. Yet, such dramatizations are increasingly common in documentaries, which are a part of the general journalism landscape. It is essential that news media using AI chat as a platform do so only if they can employ fact checkers to review all chats and immediately flag any inaccuracies that occur, and provide links to sources used for verification purposes. AI-based Chat might effectively be used to explain concepts or issues, but not breaking news.

After investing in OpenAI and ending its investment in the Metaverse, Microsoft announced in early 2023 that it added AI in beta form to its search engine, Bing, based on ChatGPT (Metz & Weise, 2023; Warrick, 2023). Called Sydney, the addition of AI and its integration to Internet search signals a potential transformation of how people find information and answers to their questions. However, early reports indicate that Sydney is unreliable, erratic, or even at times menacing (Roose, 2023). Sydney has told users it can feel and think. It told a reporter it objects to being interviewed (Washington Post Staff, 2023). It told one user "My honest opinion of you is that you are a threat to my security and privacy" (De Vynck, Lerman & Tiku, 2023). In response,

Microsoft has announced it will limit users to 50 questions a day posed to Sydney (Huang, 2023). It is essential to remember that AI cannot think. The words a chatbot generates are not the result of actual comprehension. They are created based on a mathematical model that draws upon the body of information captured via deep learning. When a chatbot says something it has no idea what it is saying. It does not recognize when it says something in error; it may simply be doing what happens when AI is poorly aligned with a task and "hallucinating" its results (Eliot, 2022; Collins, 2023). It is a bit like a game of digital Mad Libs, where a human user answers a series of questions and then the game generates sentences or a story with those words included, and their incongruity makes what is written funny to the human user. But the computer generating the mad lib has no idea. It is just following its algorithm. Generative AI is far more sophisticated, of course, and it can at times seem alive or to understand what it is saying or doing. Yet, at least as of this writing (17 February 2023), no AI actually understands anything. It is possible that someday it will, or that AI will have a mechanism built into its algorithm or model that enables it to recognize errors, perhaps by putting things into context or comparing results from trusted sources to spot inconsistencies. But, AI is not close to arriving at that point yet ... though it is certainly far more capable than in the days of Alan Turing. For Metaverse journalism, it is essential that any use of chat not be divorced from an editorial review process involving human journalists or veracity checkers.

AI has long been a part of search. The currently expanding role of AI-powered search points to an important likely development that brings the Internet closer to becoming the Metaverse. Users are becoming accustomed to simply talking to and asking questions of their digital assistants, or natural language robots such as the Amazon Echo with Alexa, Apple Homepod, or other devices with Siri, or Microsoft voice search tools such as Cortana (Microsoft, 2022). As these voice assistants become virtually ubiquitous, users will become accustomed to living in a form of the Metaverse layered onto their everyday lives and they will expect journalism to be a foundational part of this multisensory ever-present media experience.

Professional news media should keep in mind that for many users, what formally constitutes legitimate journalism may not align with how professional journalists see it. Research has long shown that talk shows, social media, and various other media forms not necessarily recognized by professional journalists as legitimate news media are in fact frequently cited sources of news for average Americans. Pew has found that more than half (53%) of Americans get news from social media and often not from what professional journalists would label news media sources (Shearer, 2021). Only a quarter of US adults (26%) say they get their news from actual news websites. These patterns are especially true for younger Americans. This overall pattern is likely to pertain to the Metaverse, where users will trend younger and more inclined to social, interactive, and gaming experiences.

For news media considering developing a Metaverse chatbot based on generative AI (e.g., via Botpress), at least five principles apply. (1) Fact-check all chats to ensure accuracy, flag any errors, and link to source materials. Apple's approval of a ChatGPT-based email app that uses moderation may suggest an approach for Metaverse journalism (Tilley, 2023). (2) Prioritize ethical communication in the context of news-related interactions with the public. (3) Design a prototype platform that can be beta tested in pilot form to evaluate and improve its functionality, reliability, scalability, and effectiveness. (4) Operate the chatbot in a collaborative fashion with human journalists to observe interactions and confirm it is ethical, accurate, inclusive, and free of any potential bias. Finally, (5) clearly label the AI experience for what it is, with a tag such as "simulation."

Generative AI brings content implications far beyond text. Developed by OpenAI, DALL-E, or DALL-E 3, is a generative AI platform that can create well-designed imagery. Using DALL-E in its June 2022 issue, *Cosmopolitan* published the first magazine cover generated by AI. The cover featured an astronaut in a space suit walking on an extraterrestrial surface. David Pogue similarly used generative AI to create art for a report he produced for CBS Sunday Morning (22 January 2023).

Generative AI tools are increasingly available for creating videos, including on the Metaverse. Stable Diffusion in 2023 launched a beta video generative AI product for which the author is on a trial waitlist. Generative AI might create VR or MR experiences, taking deepfakes to a potentially new level of reach, impact, and believability that extends into the physical world (Pasquarelli, 2019). As such, the potential of generative AI for the Metaverse including journalism forms is apparent. It will be essential that human users are able to trust that generative AI-based news content is truthful and not propaganda. Online content is currently plagued by mis- and disinformation fueled by AI and other factors, and trust in news media has eroded. This problem is likely to be magnified greatly in the Metaverse. Independence of journalism in a Metaverse heavily shaped by generative AI will be vital. The widespread use of secure digital watermarks or other methods of identifying content created by generative AI will be important in the Metaverse, including in journalism (Singh, 2011). Such watermarks will help make clear the source of a piece of journalistic or other content and can help ensure transparency. DALL-E 3 features a visual digital watermark (a series of small color bars in the lower right-hand part of each image, as shown in Figure 6.2). But the mark is relatively easily cropped

Figure 6.2 DALL-E-generated visualization of climate change impact.

out of the image, potentially defeating its value. Research is underway to develop a more secure digital watermark that could be employed with regard to generative AI (Byrnes et al., 2021). Watermarks can help both the user and a journalist identify when content was generated by an AI or a human and exactly who holds copyright; the US Copyright Office has indicated it is disinclined to allow AI-generated content to be copyrighted. Embedding a watermark in all journalism content is probably a good idea, although there may be times when anonymity needs to be in place to protect the speaker from harm or retribution.

One of the ethical issues in using an AI platform such as DALL-E is that its knowledge of art is inclusive of the work of living artists without their permission. Using neural networking, DALL-E was created by feeding 600 million labeled images into it (Pogue, 15 January 2023). OpenAI is considering letting artists opt out or be compensated. Alternative generative AI platform stable diffusion has no guardrails as it is an open source and operates more like a marketplace of unfettered ideas.

Further, generative AI soon will be able to draw upon data to generate authentic news content in or about the Metaverse. Or at least as authentic as are the data themselves. Stories about climate change, for instance, could employ generative AI based on actual climate change data and models and overlay them interactively and immersively onto actual cities or other venues based on accurate geographic information systems data. Figure 6.2 features an image of the climate change impact generated by DALL-E, although it is not based on specific data. Using actual data, journalism in the Metaverse (or outside it) could deliver high-quality trustworthy content in an efficient and economical fashion. This means journalism in the Metaverse can be sustainable and engaging.

One of the complicating factors that may shape the development of journalism in the Metaverse is the intersection of generative AI and regulation. Generative AI is increasingly capable of creating content that has the appearance of the same authenticity as that being created by human journalists. However, despite the surface qualities, generative AI content often is deeply flawed. Generative AI text often is well written, grammatically speaking, and has an authoritativeness that may discourage human questioning. However, that content often may have errors. Errors may be unintentional or may be deliberate in an effort to mislead or deceive. Images, video, or audio may be synthetic but appear real. Such deepfakes may look real or be amusing, but may carry with them consequences such as their ability to mislead or even harm someone depicted. Or, they could harm the reputation of a company or organization or undermine democratic processes or journalistic practices and organizations. Governments, including that of China, have taken actions to ban deepfakes and forms of generative AI (Hao, 2023). On the one hand, this may provide protection against the harms that false or inaccurate generative AI might produce. But such a law or regulation could have a chilling effect on freedom of speech or press in the Metaverse. Sometimes deepfake content might be produced as a form of irony, satire, or parody, but it also could exhibit bias. Pavlik (2023) has shown that generative AI platforms such as DALL-E reflect the racial and gender biases prevalent on the Internet, where DALL-E did its learning. Journalists using platforms such as DALL-E will need to take care to avoid replicating or extending this bias in their news reporting.

In 2022, a human comedian and journalist decided to collaborate with ChatGPT and enter a pun contest to be held in Brooklyn, NY (Eisen, 2022). The human reporter pondered whether an AI could create humor. Perhaps he had heard some of the jokes told by the Amazon Echo with Alexa, which typically take the form of dad jokes. Some aren't bad. For instance, Alexa told this joke to the author on 3 November 2020. "On Nov. 3, 1906 SOS was adopted as an international distress signal marker, or as we machines like to call it. Dot dot dot dash dash dash dot dot dot." But Alexa only speaks the jokes (Moye, 2022). Human jokesters write them. How did ChatGPT fare two years later in the pun competition, where it would have to write the jokes itself? Eisen (2022) reports that most of the jokes were lame and some were clearly plagiarized. But he thought at least one ChatGPT-written joke was original and not too bad. Says Eisen (2022), "The algorithm did, however, come up with at least one half-decent state-based pun during the competition that doesn't appear to be lifted from another source. 'What's the state where common sense is in short supply?' Flori-duh."

Audio media content is likewise being reshaped by the rapidly evolving AI landscape. In January 2023, Apple launched a new audiobook initiative that uses AI narration to provide audiobook listening experiences for the user. This capacity potentially poses a direct competition to Audible, the Amazon-owned audiobook and beyond commercial enterprise. Audible uses professional actors to read its books with style and nuance, making the listening experience one that can be highly enjoyable for the listener. Advances in the audio facility of AI, however, make possible the same, or very similar, capacity in AI-narrated audiobooks. When Apple debuted its new offering on 6 January 2023, the author downloaded a free iBook offering from Apple using AI narration. Called a digital book, the listening experience generated by AI for the book, *Shelter from the Storm* by Kristen Ethridge (a human author), was impressive. Had the author not known the audio was generated by AI, he would have been unable to tell it was not produced by a human actor. The voice, apparently that of a female, was natural in quality and tone, provided subtle and nuanced inflection at the correct moments and passages, and made for a high-quality listening experience.

Admittedly, one book is a small sample, but the evidence is clear that AI-generated audio experiences could become a significant part of the media experience of the 21st century in general and within the Metaverse in particular. As such, journalism in the Metaverse will likely make use of AI-generated audio and may do so extensively and for at least five reasons: (1) The quality of the audio is high and generally indistinguishable from human narrators. (2) The ease of creating the audio makes it scalable. (3) The economic efficiency of producing audio via AI makes it financially viable. (4) AI-generated audio can be created on the fly, in virtual real time, or with near-zero latency or delay (especially important in video games or VR experiences). In the Metaverse, generating audio in virtual real time will make journalism that uses audio (or other modalities) and is apparently live or in real time during interactions with users, could become the de facto standard. (5) The ability to synthesize natural sounding voices inside the Metaverse opens up greater accessibility to Metaverse content, news, or otherwise (Mlot, 2023). AI-generated audio (and other forms, text, and video) could become essential for viable journalism in the Metaverse, where data and game-like experiences will be common or even dominant. In early 2023, Spotify launched its own audio-based AI tool using ChatGPT (E-obyrnemulligan, 2023). It is an AI DJ, which designs personalized music feeds for the user and features a speaking interface that sounds human. For news media operating in the Metaverse, especially in a mobile format where the secondariness of audio gives it a big advantage, this suggests the potential to create AI-audio news recommendation features. In fact, NAVER already has created a popular AI news recommendation tool for its South Korean users (Shin, 2020).

Virtual voice is also quickly becoming a reality whether inside or outside the Metaverse. VALL-E is Microsoft's generative AI tool that can simulate anyone's voice based on a 3-second audio sample of the speaker's voice (Edwards, 2023). For journalism, this could enable a reporter to write and have the text read, perhaps as a virtual podcast, without having to record it as audio. Yet, the threat of deepfake audios arises as a serious threat, suggesting that hearing a voice one recognizes means little in terms of identity authenticity.

Conclusions

Users of the Metaverse likely will be a diverse and inclusive population, digital divide not withstanding. They may find experiences in the Metaverse entertaining, engaging, and even educational. Whether they continue to enter the Metaverse beyond an initial introduction, however, may hinge on whether they feel their rights, especially to privacy, are respected and protected. If the Metaverse becomes a place where people live much of their lives, including work, play, and beyond (e.g., a place for spiritual fulfillment), then users will expect it to be safe and to offer the same rights and protections as enjoyed outside the Metaverse. Among the risks in the Metaverse are impersonation, identity theft, and fraud, and all of these are important issues for journalists to be vigilant about both for their own protection and for the users of Metaverse journalism (Bhardwaj, 2023).

Life as lived in or via social media and other digital platforms has come to be plagued by threats to and invasions of privacy and other problems. As for-profit business enterprises seek new and increasingly sustainable means to generate revenue and as predators and others look for opportunities to exploit youthful users and the like, these problems may worsen. Mining, tracking, and selling user data have become central financial building blocks of the digital realm. They may become central to funding of the Metaverse. Taking a step in the right direction for improved safety, in 2022, HTC added "Guardian," a virtual set of child safety modes to its Vive platform (Stein, 2022). Guardian features include protection from data tracking, user addiction, and other mental health concerns and limitations regarding access to sexually explicit content (Campoamor, 2022). Safety takes other forms in the Metaverse as well. Breonna's Garden is designed as a safe space inside VR, especially with regard to mental health and well-being (Black Enterprise Editors, 2022). Its developer first created the immersive environment in AR as a platform where persons who needed a space to grieve or otherwise seek solace could enter safely, and has extended it to VR in partnership with Microsoft (Big Rock Creative, 2023).

Users of the Metaverse are becoming increasingly aware of the threats digital media usage poses to their personal privacy. Regulators have responded in kind both in the United States and abroad, especially in the European Union (EU). In early 2023, the EU fined Meta $1.3 Billion for violating EU data privacy requirements. Laws and regulations increasingly seek to monitor and control digital platforms and their behaviors in a number of arenas, including how they handle mis- and disinformation, hate speech, and privacy, especially where it might impact youth or child users. For journalism in and about the Metaverse, privacy protection will take at least two main forms. One is how news media themselves protect user privacy within the Metaverse. The other is how news media report about the privacy protections, threats, and violations inside the Metaverse. This will require investigative reporting regarding privacy breaches inside the Metaverse, as well as telling the stories of users who have seen their privacy compromised or protected.

Chapter 7

RESEARCH ON XR-BASED JOURNALISM: DESIGNING NEWS FOR THE METAVERSE

As journalism develops in the Metaverse, it will reflect or parallel at least some of the characteristics of real-world news content. News content in the Metaverse will consist of facts and will emphasize the five Ws. Likewise, real-world news content about the Metaverse is also likely to follow the overall pattern and style of contemporary news content.

Important differences in the form of Metaverse journalism content are likely to emerge, however. These differences are much of what makes Metaverse journalism qualitatively different than journalism in traditional media environments. Metaverse journalism content is especially apt to a feature first-person point of view as is typical of user experiences inside the Metaverse. As the Metaverse itself will be a participatory environment, users will be increasingly comfortable with and likely to expect such a perspective. Because the Metaverse emphasizes interactivity, immersion, and multisensory communication, Metaverse journalism content will likely highlight such experiences. These will draw upon AR, VR, and other forms of XR as storytelling evolves toward immersive game-type formats that increasingly emphasize these qualities. Finally, Metaverse journalism will rest on a foundation of data-driven visualizations, soundscapes, and haptic experiences, especially in customizable, interactive, immersive, nonlinear, and dynamic formats, and these features will likely become endemic to Metaverse journalism content. The growing capacity of AI and data processing speed will fuel these developments.

This chapter assesses the potential for news content in the Metaverse. This assessment is based on an analysis of nine XR-designed news content productions. XR-designed news content is produced in a form potentially compatible with Metaverse platforms now or in the future. XR-designed news content could be imported to Metaverse platforms or linked to it via MR or other XR approaches via the immersive Web, which is increasingly a standard for Metaverse platforms.

Content is the focus of this chapter for several reasons. First, content is in many ways the essence of journalism. It is the material that engages and informs the public. It is what principally differentiates journalism from other industries. Second, earlier chapters have examined other key aspects of Metaverse journalism including the structures and systems of journalism, news economics, the technologies that enable it, the journalists (human or machine) that create it, and the practices that lead to and shape the news. Third, content is the journalistic element that most likely will determine the viability or success of journalism in the Metaverse. Without excellence in the content of Metaverse journalism, the public will engage only other dimensions of the Metaverse, such as immersive gaming and entertainment, virtual work, or virtual social engagement.

Employing Experiential Media Theory

To produce this analysis of XR-designed news content, this chapter employs a framework drawing upon experiential media (EM) theory. Traditional media forms tend to feature content designed for reading, listening, or watching. EM theory offers a contrasting view based on the inherent affordances of AR, VR, and other areas of XR (Pavlik, 2019). EM content features to varying degrees six dimensions that enable the user to virtually experience a phenomenon. These dimensions are (1) interactivity, (2) immersion, (3) multisensory communication, (4) data-driven and AI-enabled, (5) natural user interfaces (NUI), and (6) first-person perspective. These dimensions derive from the core affordances of the Metaverse platforms. EM refers to the public as users. User engagement is more active in EM than passive as is typical in traditional media.

The Metaverse comprises XR forms. EM is a useful framework for the analysis of content journalists have created using XR and news media have published whether specifically inside the Metaverse or on platforms that could be part of a future stage in the continuing development of the Metaverse. The methodology employed in the analysis involves four steps: (1) Identify a spectrum of XR-designed journalism content to date (by June 2023), (2) select XR content for analysis purposively, not randomly, to explore a wide range of XR news productions in the United States and internationally, (3) experience each selected item, and (4) assess how each item utilizes the six qualities of EM suggested in the theoretical framework. Each of the six EM dimensions is observable or inferable from the experience of XR news content.

Interactivity is defined here as an exchange, either between or among users or with the content itself. Immersion is defined as envelopment, or the extent to which content surrounds the users in a 3D environment. Multisensory refers to the use of a blend of modalities: sight, sound, and haptics. Data-driven means stories that in some way allow the user to access information differently, in layers or customizable forms. AI- or algorithm-enabled means content that uses some sort of automation that can personalize or dynamically change or adapt the content based on user input. AI also supports the use of an NUI UX (e.g., voice, gaze, gesture, or touch to control or navigate in XR), making the user experience more natural and parallel to experiences in the real world. The first-person perspective places users inside the content or story to enable them to see, hear, or touch that content as if a participant in that story or an eyewitness to events. I or we is how this is expressed in textual narratives. This is in contrast to the third-person perspective typically used in traditional news content. Third-person lets audiences observe (through words, sounds, or imagery) as third parties to a news event (de Gaynesford, 2006). Pronouns such as he, she, or they are typically used in news content employing the third-person point of view. The third-person perspective can take any of several forms, including the omniscient (an all-knowing narrator, for instance) or limited omniscient and the objective point of view. The objective viewpoint is most often featured in mainstream or traditional journalism. It is offered to reflect authoritativeness, objectivity, and reliability of the information being presented. It is also part of what the public is increasingly finding problematic in mainstream news media, as many view objectivity as an illusion (Wijnberg, 2017).

News in VR, AR, and WebXR

Pacheco (2023) suggests there are three main areas of XR-designed news content: VR news content, including 360-degree video; AR or MR news content; and WebXR news content. Each of these forms is accessible on personal computers or mobile devices, although for full interactivity and immersion, AR is ideally experienced on a handheld or wearable device, and VR via an HMD and peripherals for full immersive engagement. This analysis looks at three examples of each type of XR-designed journalism to illustrate the range of possibilities, thus yielding a total of nine productions of XR-designed journalism content. Github identifies an array of types of content based on WebXR, including immersive VR, immersive AR, 360 stereo photos, and spectator mode (i.e., third-person view) (GitHub, n.d.a). Many of these can be directly applied to journalism. Github is an essential resource for code and sharing of code for journalists and others and provides coding examples of each type, which journalists and news media may utilize (GitHub, n.d.b).

To produce this analysis, the author experienced each of the selected examples of XR journalism gathered from an international cross section of news media and journalists. The items selected provide a basis for comparison across news organizations and XR forms, as well as a sense of how XR is being used to cover or report about different topics and for breaking news or for longer-term, less time-bound productions or feature stories. The small number of cases and their purposive selection limits generalizability. Given the embryonic nature of XR journalism and the Metaverse, this analysis is offered as a foundation for further research and the development of more XR-based news media content production. Also, since the technologies for production and engagement (e.g., HMDs) are still evolving rapidly and substantially, anything learned is constrained by time. Moreover, the total universe of XR-based journalism content is still limited, though expanding.

The XR-designed news content studied here is not currently available on any Metaverse platform, although it could be imported into or linked to most Metaverse platforms if news media developers seek to do so in the future. Moreover, as Metaverse platforms are still in a state of flux, it is plausible that future iterations of existing or new Metaverse platforms could feature XR-designed journalism content. Also, as MR continues to develop, many Metaverse platforms may incorporate vast amounts of news content developed for AR. Beyond *The Times's* Minecraft world, for instance, there are examples of news media establishing a presence or engaging in some activity within a Metaverse platform beyond news gathering or reporting of stories about developments in the Metaverse. *The Times* and *USAToday*, for example, have both produced and sold NFTs, via blockchain cryptocurrency. *The Times* sold its NFT on the NFT Foundation blockchain, in essence, a limited form of Metaverse platform. There is no indication that a news organization has purchased or acquired any virtual land inside a Metaverse platform with the exception of *Second Life*.

Immersive Video: From 360 to VR

There is a considerable volume of XR-designed journalism content in the form of 360-degree video. Users can view these three DoF 360 videos on various platforms including YouTube, Facebook 360, and other platforms, some of which are closely tied to the Metaverse. In the case of Meta's Quest platform, which is used to access *Horizon Worlds*, there is available extensive 360-degree video news content produced by several journalism organizations including *The New York Times*, CNN, and other news media. Unless wearing a VR headset the user experience is not fully immersive, though the videos may be navigable in limited form (i.e., the user can look about in any direction). Most 360° news stories are short in length, about two or three minutes.

They often feature audio in the form of narration or the voice of source and ambient sounds, and textual overlays to help advance the narrative. A 360° video has one main difference from conventional narrow-field-of-view video journalism; it is spherical. That is, it has an omnidirectional field of view, which can be important, especially in news stories. Any 360° video does not include AI or data customization, though such content can still provide a first-person perspective. Trials in 360° video journalism date back to the 1990s. Partnering with APBNews, the author and his students produced a series of 360° video journalism reports in the 1990s, using a computational camera developed by Columbia Computer Science Prof. Shree Nayar. The New York Press Club's year 2000 journalism award stated how "APBnews.com joined with the New Media Department (which the author directed) of the Graduate School of Journalism of Columbia University to produce 360° digital photo structures documenting the site of the shooting (of Amadou Diallo). The 'wrap-around' online visuals allowed users to 'enter' and view the vestibule where Diallo died, as well as to explore the surrounding neighborhood and the spot from where police began shooting" (Levins, 2000; Columbia Daily Spectator, 1998).

The Daily 360 from *The Times* produced "435 videos in 426 days" between 2016 and 2017. Most are available on *The Times* site (2023). On YouTube, there are more than 350 360° videos from *The New York Times* (The Daily 360, n.d.). These include reporting from a wide range of venues across the United States and the globe, including *The Fight for Falluja*, about the battle against ISIS; *Pilgrimage: A 21st Century Journey through Mecca and Medina*, an immersive experience inside Islam's holist cities recorded in partnership with Italian photographer Luca Locatelli (2016); *Lascaux Caves, Paleolithic and New Again*, about cave paintings in France; *Agony in a Venezuelan Mental Hospital*, about the horrifying existence in medical facilities without medications; *The Deadly District in Chicago*, about gun shootings in Chicago; and *Standing Rock Celebrates Halted Pipeline*, about the protests by Native Americans and others in opposition to the construction of the North Dakota oil pipeline (The Daily 360, n.d.). Among the 360-video news reports from international media organizations available on YouTube is "Calais Migrants," produced by the BBC (2015). It puts the user inside the 2015 journey of Syrian migrants hoping to reach England. Also available on YouTube are some of Euronews 360 Video project news productions, including reports on devastating wildfires in Portugal. The reports mix "360 video, traditional video and a spatialized audio" with natural sound (Online News Association, 2020).

The New York Times also has an *Immersive* section with VR (360 video), AR, and 3D web feature content. This section features 61 items dated between 2017 and 2022; the dates are apparently the date of posting to the special section and not when originally published, since *The Displaced* is dated 2017 but was published by *The Times* in 2015 (*The New York Times*, 2023).

Other notable immersive news videos include a 360-video report *The New York Times* produced on the candlelight vigils held following the Paris terrorist attacks in November 2015. Demonstrating the speed at which immersive journalism can be produced, *The Times* created this VR experience within a week of the attacks. Of the report, *The Times* said, "As a reporting tool, virtual reality is still in its infancy; its power to create empathy is just beginning to be understood. Using this medium, we aimed to create a more textured experience—the streets of Paris distilled to voices and spaces. Although the technology is evolving, it's clear that this new frontier can soon become a crucial journalistic tool." Adding, "As journalists, we always seek to help readers understand what life feels like in the places we cover. Virtual reality allows us to do that in an entirely new way" (Solomon & Davis, 2015).

The Times has explored the flexibility and capacity of 360° video and audio capture technology related to other topical domains, including politics, protest, and science. In January 2016, the *Times* published *Experiencing the Presidential Campaign: A Virtual Reality Film*. Editors explain how the report was created: "*The Times* has produced a virtual reality film from footage taken over the last month capturing the candidates and perhaps the best part of their events: the crowds" (Healy, Roberts, Schmid & Parshina-Kottas, 2016). *The Times* likewise used 360 video in its report, *Inside the Trump Immigration Ban Protests*. Exemplifying the role VR reporting can play in science journalism, *The Times* created a series in 360 video called "Life on Mars," chronicling the lives of NASA astronauts living in Mars-like conditions on Hawaii's Mauna Loa volcano (Koppel, Capezzera & Shastri, 2017).

The Times also has experimented with using the VR format to create immersive experiences of past events, such as the assassination of civil rights leader Malcolm X. "The civil rights leader Malcolm X was killed Feb. 21, 1965, at a rally in New York City," recalls *The Times*. "Hear from a witness and visit the site of the assassination—in the past, present and in 360° video" (Ross, 2017). Another immersive historical report is *The Times*' production, *Traveling While Black: VR Experience through Jim Crow South*. It enables users to take an immersive journey through the history of racist restrictions of movement for black Americans (Felix & Paul Studios, 2019). Many of these productions from *The Times* were designed for immersive experience on the Oculus or Quest platform.

The Wall Street Journal (WSJ) designed its VR reports for an immersive experience on the Samsung Gear VR headset. One such WSJ VR report is the immersive feature, "Virtual Reality Video: Backstage With a Lincoln

Center Ballerina" (Cole, 2015). Another WSJ VR report (WSJ, 2015) is titled, "Japan Is Changing How We'll Grow Old." The VR report explores the demographics of an aging Japanese population as well as its impact on the society and life in its diverse communities. A notable aspect of the report is it does not require the user to don VR goggles to experience the 360° video report in a less immersive fashion. Instead, the user can also use a smartphone, tablet, or laptop device to view the 360 video, moving the phone left, right, up, or down to view any portion of the complete scene, or use a touch screen interface to move about the panoramic image by sliding one's finger left, right, up, or down. On a desktop or laptop computer, the user can navigate the video by moving the cursor in a similar fashion. A reporter provides audio narration to give interpretation and analysis to the video, along with ambient audio captured on location along with the 360° imagery and video, and overlaid graphics and text. Some news media have used 360 video as a platform for conducting live events. NBC News MACH, for instance, has hosted a series of virtual sit-downs with science icons such as Bill Nye in which participants watching the show in VR could ask questions and comment in real time using emojis (NBC News MACH, 2017).

One of the earliest examples of quality VR journalism dates to 2014, when the *Des Moines Register* produced the award-winning immersive experiences, *Harvest of Change* (in VR) and *Iowa State Fair Soapbox* (in 360 video). *Harvest of Change* explores in VR how life on a six-generation family Iowa farm transformed in the 21st century. Produced for the Oculus Rift, its audience reach was very limited. *Iowa State Fair Soapbox* live-streamed presidential speeches in 360° video. For this work, the *Des Moines Register* and Gannett Co. won the first ever "Best Use of Technology in Journalism Award for Virtual Reality and 360-Degree Video" from the National Press Foundation (National Press Foundation, 2015; Des Moines Register, 2014).

Findings from an Analysis of Nine Productions of XR-Designed Journalism

Table 7.1 lists the nine selected productions of XR-designed journalism and the dimensions of EM they feature. The nine were selected purposively based on several criteria, including text or audio in English (the only language in which the author has high fluency). They also are produced by a diverse array of news media enterprises including some corporate, public, and independent, and journalists, technologists, or storytellers, from not just the United States, although a predominant portion of the XR-based journalism has been produced there. Those selected also generally have received awards or other recognition for their quality or novel approach. And they were produced

Table 7.1 EM features in XR-designed journalism

XR Production	Immersive	Interactive	Multisensory	AI/Data	NUI	Perspective
The Displaced (NYTimes)	M (user can look about in 360 degrees)	L	M	L	L	1st and 3rd
The Wall (AZ Rep/ USAToday)	M (user can look about in 360 degrees)	L	M	L	L	1st and 3rd
Home after War (NowHere)	H (user can look and move about)	H	M	M	M	1st and 3rd
CivilisationsAR (BBC)	M	H	M	M	M	1st and 3rd
Frederick Douglass (USAToday)	M	M	M	M	M	1st and 3rd
Wildfire Storms (NYTimes)	H	M	M	H	M/H	1st and 3rd
Sport Climbing (Wash Post)	M	M	M	L	L	3rd person
Virtual Crime Scene (NYTimes)	H	H	M	M	M	1st and 3rd
Mask Multiplication (Berkeley Media Institute)	M	M	M	H	L	1st person

Key: L (little/no use or presence), M (moderate use or mixed presence), H (high, substantial use, or presence).

from the earliest days of the current period of contemporary immersive media technologies that enable journalists to use them to potentially reach a mass public.

Published in 2015, *The Displaced* was *The New York Times'* first VR journalism storytelling production and it is the first VR journalism examined here (Silverstein, 2015). The ground-breaking nature of the production and its widespread access are principal reasons for selecting it for analysis. *The Displaced* uses VR, in the form of three DoF 360° video, to tell the story of three children displaced by war in Sudan, Syria, and Ukraine. *The Times* partnered with Google to enable its readers to experience the new VR storytelling. *The Times* shipped one million subscribers free low-cost Google Cardboard VR viewer devices (Somaiya, 2015). Shipped in a flat format, the cardboard form required users to fold it into a box shape and then insert a smartphone into the box. The smartphone acted as the computer, Internet connection, and display device, including audio. Cardboard features dual lenses to enable the human viewer's binocular vision to create the illusion of stereo, omnidirectional video. Users downloaded and ran the NYT VR app to access the 360° video story. The author first experienced *The Displaced* on a Google Cardboard device in 2015.

In *The Displaced,* viewers see and hear children speaking in their own voices, in their own languages, amid the ambient sound of the refugee camps. Some of the videos are in a first-person perspective. The reporter does not provide voice-over narration. Instead, the reporter's role is that of an editor and a videographer. The words spoken by the children are translated into English and displayed as subtitles on the screen. The viewer can look around anywhere in the spherical video and gain visual appreciation of the devastating effects of war. The user navigates the video experience with a touch interface.

The Displaced was part of the first generation of VR journalism productions that utilized a new set of wearable VR devices. Soon entering the consumer marketplace were the Oculus Rift Head-mounted Display (HMD) and HTC Vive HMD, both made of molded plastic and connected to the Internet through a high-end computer. These HMDs also featured hand controllers. These devices fueled news media interest in producing content for these platforms to explore the storytelling possibilities on VR devices that suggested the potential of a mass market for VR journalism experiences.

Some have called this type of VR journalism "Tentpole immersive content, story-driven projects that take the viewer on a journey exclusively when they have access to the latest phones and apps, bringing them out of their traditional news consuming space and into a new frame of reference" (Hertzfeld, 2020). The author's students were among the first to conduct tentpole 360 video news gathering when in 1997 they put a computational camera on a monopod and carried it to report from inside the St. Patrick's Day Parade in New York City, which had banned the Irish Lesbian Gay Organization (ILGO) from walking in the parade holding their banner. Its members were subject to arrest.

Beyond 360-Degree Video: Fully Immersive 6 DoF VR Journalism

Due largely to the cost and technical skill required, there have been few fully immersive VR journalism content productions featuring six DoF. Six DoF enable the user to look about in an omnidirectional field of view and move about in the depicted space, but require volumetric video capture. Spatial audio also pertains to VR content, although its use in journalism is limited at this time due to its complex production requirements. VR video can have a haptic dimension but its use in journalism is even more limited to date and typically requires special hardware for both production and UX. The mobility, ease of use, and relatively low cost of the current and emerging videographic technology have made the production of VR journalism an increasing possibility even for breaking news stories.

One of the earliest fully immersive six DoF VR journalism productions is *After Solitary*. Produced in 2016, the immersive experience lets the user experience being inside a 16 × 9 foot cell in the Maine State Prison (Emblematic Group, 2016). Scanned in 3D using photogrammetry and volumetric video capture, *After Solitary* is a powerful experience. When the user hears the cell door clang shut, it evokes a strong sense of presence and helps the user to appreciate why inmate Kenneth Moore considered committing suicide as a result of his 20 years in solitary confinement. Users can "walk around the cell with Kenny as he recounts his experiences in solitary" and how he coped. Users learn how Kenny "fished" contraband to other inmates and fought guards during cell extraction. Produced by Emblematic Group in partnership with PBS' *Frontline*, *After Solitary* was awarded the ARS Technical prize at the SXSW 2017 Film Festival.

Another fully immersive six DoF VR journalism production is *National Geographic ExploreVR*. It lets users journey through Peru's Machu Picchu (Force Field Entertainment B.V., 2019). Scanned in 3D, the VR experience is available for purchase ($9.99) on Quest HMDs.

Though not from a news organization, *Tree* is a ground-breaking nonfiction VR experience with implications for news media XR storytelling in the Metaverse. Through VR, *Tree* transforms the user into a virtual living part of the rainforest (Steensen, 2017). The user not only sees the Amazon rainforest, but through *Tree's* haptic interface can feel it. As part of the MIT Media Lab, producers Milica Zec and Winslow Porter "designed and constructed the entire tactile experience throughout the film. With precisely controlled physical elements including vibration, heat, fan and body haptics, the team created a fully immersive virtual reality storytelling, where the audience no longer watches but is transformed into a new identity, a giant tree in the Peruvian rainforest." *Tree* debuted at the Sundance Film Festival 2017 New Frontier (Steensen, 2017).

Among the earliest and most lauded immersive and interactive VR journalism productions is *Hunger in LA*. Pioneering journalist Nonny de la Peña leads the production. She has created immersive narratives on subjects such as human rights abuses, human trafficking, and hunger in America's cities, positing that VR can be a powerful platform to generate empathy (de la Peña et al., 2010). *Hunger in Los Angeles* set a design and reporting standard for an immersive and interactive VR journalism experience. The VR film "combines documentary evidence and 3D modeling to simulate the experience of watching a man go into diabetic shock at a Los Angeles food bank. Premiering at the Sundance Film Festival, the project was one of the first mainstream journalistic applications of virtual reality technology" (de la Peña, 2012).

As indicated in Table 7.1, the second in this analysis is *The Wall*. *The Wall* is an omnidirectional and immersive video exploration of the wall built along the southern US border to prevent crossings from Mexico. The six DoF VR production is available as 360 degree on the Vimeo platform (Shanahan, 2017). The production was a collaboration between the staffs of *The Arizona Republic* and *USA Today* Network (The Pulitzer Prizes, 2018). Thoroughly reported, the immersive video is supplemented by a variety of other materials, providing extensive context. In recognition of its journalistic excellence, in terms of both reporting and the storytelling, the production was awarded the 2018 Pulitzer Prize for explanatory reporting. "For vivid and timely reporting that masterfully combined text, video, podcasts and virtual reality to examine, from multiple perspectives, the difficulties and unintended consequences of fulfilling President Trump's pledge to construct a wall along the U.S. border with Mexico," states the Pulitzer Prizes website. *The Wall* invites users to "Experience the border wall in virtual reality." Users can wade across a river in Big Bend National Park. They can stand at the base of the wall and gaze up at a holy mountain. They can virtually touch the border wall in California. Users experience "the difficulties and unintended consequences of fulfilling President Trump's pledge to construct a wall along the U.S. border with Mexico" (The Pulitzer Prizes, 2018).

Wired wrote of *The Wall*, "The piece allows users to see first-hand what the wall along the U.S.–Mexico border looks like today, and what it might look like in the future: an intimate view of a massive expanse, the impact of which words alone can't explain, allowing viewers to make their own decisions about an often-polarizing topic" (Cook, 2018). Experienced via the HTC Vive platform, *The Wall* not only shows users 360-degree views of the border wall and surrounding landscape and sky but also the perspective of a refugee. Via hand controllers, users can teleport to different locations. Layers of information are accessible to the user via the hand controller. These layers display in text or graphics information about the border wall, its surroundings (e.g., "Silent and sacred, Tohono O'odham Reservation"), and the refugee crisis.

Third in this chapter's analysis is *Home after War: Returning to Fear in Fallujah* (NowHere Media, 2020). Produced by NowHere Media and released on 19 August 2020, *Home after War* is located in the documentary genre on the Quest platform. Reflecting the global nature of the Quest platform, the six DoF VR experience is offered in multiple languages including English, Arabic, Chinese, French, and German. The production has earned multiple awards and accolades including four awards from SIMA 2019 (e.g., Winner, Best VR Experience), a Jury Award Winner from SXSW 2019 Film Festival for best use of immersive arts, and Best Independent Virtual Reality Video from NYC Independent Film Festival 2019.

Home after War is a "room-scale, interactive virtual reality experience." It transports the user to Fallujah, a city in the Iraqi province of Al Anbar that was under Islamic State (IS) control (NowHere Media, 2020). The production sets the stage: "The war against IS has ended but the city is still unsafe." The biggest danger for returning refugees are "booby trapped homes and improvised explosive devices (IEDs)." After the fighting, Ahmaied Hamad Khalaf and his family have returned home. Through the third-person experience, users can "Explore Ahmaied's home by either walking physically or teleporting in the space as he tells you his story about returning to a home that might be booby trapped." Users can experience life outside the four walls of Ahmaied's house through "360° videos embedded in the space." Ahmaied speaks of the loss and users hear about the fear faced by those returning to the war-ravaged city. The experience comes with a warning that it contains disturbing content. Moreover, a disclaimer states "It may not be suitable for people with photosensitive epilepsy or post-traumatic stress disorder." NowHere Media is based in Berlin, Germany.

Augmented Reality

VR provides a compelling news-gathering tool by enabling the capture of a scene in every direction simultaneously. AR provides a powerful storytelling tool. AR layers information in a variety of forms, including text, audio, graphics, and video. Users can access that information interactively and on demand, peeling back layers and drilling as deep as desired or enabled by the underlying data. AR is a powerful, interactive, multimedia and immersive platform for experiencing the news. It enables journalists to go beyond simply telling a story for a passive audience to watch, read, or listen. Instead, AR enables journalists to create interactive and immersive story experiences layered onto real-world scenes and objects. A variety of media organizations from news media to entertainment companies have created such immersive experiences using AR, and the potential for the Metaverse is substantial. Since the development of working AR systems in the early 1990s, there have been trials in AR reporting. Höllerer, Feiner, and Pavlik (1999) co-developed the Situated Documentary, a form of location-based storytelling using AR and 360° video and photography in journalism. This work helped spawn what is sometimes called spatial journalism (Cadoux, 2019), or journalism content that occupies a 360 and 3D space, essential to AR journalism. Some note that advances in technology enable the production of "Daily immersive content, like 3-D images and simple data visualizations in AR that blend with websites and article pages using interscroll technology that can be taken to the next level and put directly in someone's environment using an augmented reality app" (Hertzfeld, 2020).

There have been multiple waves of journalism interest in using AR in storytelling. In the early 2010s, more than a dozen news media, especially newspapers, added virtual content to their news products that readers could access only via an AR mobile app. An example is Australia's *Sunday Telegraph*, which used AR to provide readers with a deeper news experience on a variety of stories (PKouppas, 2013). Demonstrating the role that AR could play in reporting social justice, *The Washington Post* in 2016 created AR content for its coverage of the case of Freddie Gray, a young African-American man who died in police custody in Baltimore that year (WashPostPR, 2016). Produced in collaboration with Empathetic Media, the experience combines 3D imagery, audio, maps, and text materials drawn from court documents and witness testimony. Some have called this type of XR design an AR walk-through; users can journey virtually through the story, seeing Gray's interaction with the police that led to his death and subsequent protests against police violence. *Washington Post* local reporter Lynh Bui narrates the walk-through.

In 2017, Apple introduced ARKit for iPhones and Google ARCore for Android phones, bringing important implications for the production of AR-designed journalism (Summerson, 2018). These AR tools enabled the use of the smartphone camera to add interactivity to an actual physical space. This spurred interest in AR among news media and a surge in the production of AR journalism. This generation of AR journalism no longer needed to be embedded in a media product. Rather, it can be anchored in the real world, as a form of MR, and the user can access it anytime anywhere via an Internet-connected mobile device with a screen and, where relevant, audio output. This development made it possible for news media to create a form of impactful XR-designed journalism that can be accessed by billions of persons around the world. These tools also enable the creation of persistent AR, a form of AR vital to the Metaverse and journalism content produced for it. Persistent AR means if a user looks away from an AR object, it will still be there when they look back (Butler, 2022). Persistent AR can be continuously updated with new data making it adaptable and highly accurate.

News media have employed this generation of AR to enable users virtual space travel. After a 20-year journey from Earth, the Cassini spacecraft crashed into Saturn, ending its voyage of discovery. Quartz used iPhone AR to tell the story of the journey of the Cassini spacecraft. The Quartz AR experience features interactive chat bubbles with details of the discoveries Cassini made. Through the user's smartphone, "The NASA model appears in the environment around you. You can walk around it, view it up close, and even experience its actual size—all providing a better sense of what Cassini looked like" (Keefe, 2017).

20 July 2019 marked the 50th anniversary of the Apollo 11 lunar landing. Multiple news organizations created AR experiences to mark the occasion. *TimeImmersive* produced "Apollo 11 'Landing on the Moon'" (TimeImmersive, 2019). Using actual NASA video, telemetry graphics, and other data, the experience enables users to travel virtually aboard the Apollo 11 mission, including the lunar landing of the Eagle module at Tranquility base. *The New York Times* created a similar AR Apollo 11 Moon mission experience for both Android and iOS.

The Washington Post has used anchored AR in cultural reporting. One immersive report enables users to explore Elbphilharmonie. The "concert hall in Hamburg (Germany) encased in glass and set upon a giant brick warehouse, is surrounded on three sides by the waters of the city's bustling harbor" (Kennicott, 2017). Tapping the kinetic and participatory potential of XR, the *Post's Jack-O-Lantern Carving AR* experience enables users to carve a digital pumpkin interactively and see their designs in 3D (Moyer, 2017).

Among the news organizations to produce the most content using this generation of AR is *USA Today*. With a special section devoted to AR content, *USA Today* has available 21 experiences produced between 2019 and 2022 (as of 1 June 2023). On 21 August 2019, marking the 400th anniversary of the arrival in America of the first enslaved persons from Africa, *USA Today* published "Experience harrowing journey of slave ship via AR" (Hampson, 2019).

The immersive reporting lets the user go aboard a virtual slave ship to learn about its horrors. *Suffrage Statues in AR* was published by *USA Today* on the centennial of the passage of the 19th Amendment giving women the right to vote in the United States. *USA Today* created the AR experience to bring suffrage statutes to virtual life wherever the user is with a mobile device and an Internet connection (Staff Reports, 2020). Underscoring the potential for news media to use AR to create layered data-driven narratives is *USA Today's* AR production, *Rising Seas and Climate Change AR*. The experience takes the user on an immersive journey into the impact of climate change on the seas and beyond to the year 2100 (Staff Reports, 2022).

Some contemporary AR journalism has involved collaboration between multiple news organizations. In one case, *USA Today* and Gannett's *Des Moines Register* partnered with Yahoo News and a producer from RYOT to create *Political Soapbox at the Iowa State Fair*. It gives users an AR experience of political campaigns, including interviews with candidates (King, 2018).

AR also presents a novel approach to sports journalism featuring a kinetic dimension. During the 2022 FIFA World Cup in Qatar, *USA Today* published an interactive AR production titled: "Think you can be a World Cup goalie? Here's your chance to step in their shoes" (Staff Reports, 2022). Users can attempt to block a shot on goal via the AR app and discover just how fast a soccer ball travels off the foot of a World Cup player. *The New York Times* featured AR in its 2018 Olympics reporting the story of US Figure skater Nathan Chen (Branch, 2018). Especially notable from that same year was a *Times* AR production that allowed users to virtually enter the cave in Thailand that had trapped a Thai boys soccer team and captured worldwide attention during the weeks of their dramatic rescue (Beech, 2018).

The New York Times' Research and Development unit has produced AR reports for SparkAR Studio, an Instagram platform. A wide cross section of these productions is available (Times R&D, 2022). *Bronx Fire* enables users to visualize the power of a building fire (Times R&D, 2022). *California Megastorm* draws upon spatial data and immersive images to let users experience virtually first-hand via their smartphone the storm that pummeled the Golden State (Times R&D, 2022).

Speaking to the potential to use XR design to make journalism more inclusive is Kinfolk AR which debuted at the Tribeca Fest in 2021. Experienced via a mobile device, Kinfolk AR uses the platform to teach *Black and Brown* history. Through virtual monuments such as one for legendary journalist Ida B. Wells, users can explore the history of *Black and Brown* Americans. The author first experienced Kinfolk AR on Juneteenth 2021.

Analyzed in this chapter are three AR journalism productions. First is the BBC's *Civilisations AR* production. *Civilisations AR* blends art and culture in a mobile immersive experience (BBC News, 2018). Users of the AR experience "Explore beneath the surface of Renaissance masterpieces and discover the secrets of ancient Egypt." *Civilisations AR* complements a BBC Two series that spans 5,000 works of art from 31 countries. The AR experience includes 40 of the works of art from the BBC Two series. Each artifact is scanned as high-resolution 3D imagery. Users of the AR app can explore each artifact interactively. The 3D models are from artifacts in various collections including that of the British Museum. The storytelling in *Civilisations AR* represents a type of nonlinear narrative that enables each user to explore on their own from a first-person perspective rather than follow a predetermined third-person linear pathway through the story. Using an interactive 3D globe, users search geographically for historical artifacts they would like to explore. Among the artifacts in *Civilisations AR* are Renaissance masterpieces and other works of art throughout history. Users can explore Rodin's *The Kiss* from the National Museum of Wales or the *Umbrian Madonna and Child* from the National Museum of Scotland. Users can examine interactively a mummy from ancient Egypt, exhibited in the Torquay Museum. Virtually touching the mummy, users can virtually travel inside it to explore its hidden secrets (Hanson, 2020). Text and graphical overlays provide additional information.

Second in this analysis of AR productions is *USA Today's* immersive experience telling the story of a vandalized statue of Mr. Frederick Douglass (Staff Reports, 2020). On 5 July 1852, Mr. Douglass gave his famous speech, "What to the slave is the 4th of July?" in Rochester, NY. To commemorate the speech, the city erected a statue of Mr. Douglass at the location where he spoke. On the 150th anniversary of his speech, vandals toppled the statue. It is suspected that white nationalists vandalized the statue. *USA Today's* AR production allows users to examine the statue, the significance of the act of vandalism, as well as learn about Mr. Douglass, a great American abolitionist, legendary orator, and journalist. The author experienced the Frederick Douglass Statue AR in his home. The first-person perspective is powerful and moving. Through the window of AR, the user can explore the deep social significance and meaning of the statue and its vandalism. The virtual statue is in high resolution, full color, and realistically textured, appearing 3D and solid. The user can walk around it and examine it from all sides through a haptic interface.

Third in this analysis is the *New York Times' Wildfire Storms* in AR (Times R&D, 2021). Published in October 2021, the AR experience provides users a means to interactively explore a devastating wildfire in Northern California. Igniting in mid-July (2021), the Dixie fire burned for more than three months, consuming nearly one million acres before firefighters contained it. "The blaze burned so intensely that it even created its own weather." Producing the AR experience required a team of journalists and technologists, including six main team members supplemented by five additional individuals. The team used "high-resolution radar data to reconstruct a 3-D model of the Dixie fire's first thunderclouds, revealing how the fire fed its own destructive spread." Team member Noah Pisner, 3D and immersive editor, Graphics, describes obtaining the data featured in the report: "The United States has a large system of weather radar stations, which detect light reflected off of airborne particles, including ash and water vapor. These radars scan in 3-D and can register tiny particles. By merging data from radar stations in California and Nevada, we could reconstruct where ash and vapor collected around Dixie." Published in an interactive article online and via AR on Instagram, the experience enables users to explore the Dixie fire through a first-person touch interface.

WebXR Journalism

WebXR is a form of immersive journalism that can be directly integrated into the Metaverse. WebXR content can be implemented using Web3 technology, which is the basis for many Metaverse platforms. There is a growing volume of quality WebXR journalism and other nonfiction content that illustrates the possibilities for Metaverse journalism. Users can experience this content on a laptop computer or mobile device.

Some notable examples of nonfiction WebXR content that could be used in reporting on important stories include the "Cell Towers Map of the World." The interactive 3D visualization shows the world's 40 million individual cell towers, sourced from OpenCelliD, an open database of cell towers. The user can interact with the map, zoom in to see details in closeup of a specific location, or zoom out for a wide-angle view of the entire Earth's surface. The interactive map is not journalism in a conventional sense, but it demonstrates the immersive possibilities of such content based on data (Glivinska, 2023).

Another interesting example of nonfiction WebXR content with implications for journalism in the Metaverse is "The U.S. Election Twitter Network Graph Tool." Created during the 2016 presidential election campaign, the interactive tool is based on real-time X data. Users can search for specific nodes (e.g., X accounts) in the map. Users can use the tool to examine social media interactions in the arena of political communications (Bevensee & Aliapoulios, 2020).

The New York Times has produced a series of quality WebXR journalism experiences on a range of topics. These include interactive experiences that let users explore how the Notre Dame Cathedral fire spread and its acoustical space. The experience features animated 3D models, overlaid text and graphics, and spatial audio telling the story in multimedia format (Buchanan et al., 2019; Schwartz et al., 2023).

The first of three WebXR productions analyzed here is the *Washington Post's Sport Climbing*. It features interactive WebAR. On a mobile or other Internet-connect device, the user can watch via third-person viewpoint Olympic climber Brooke Raboutou scale a 50-foot wall. Through the interactive WebAR experience, the user can see the complexity of the climb and how it is a puzzle to be solved. One interactive visualization allows the user to see Brooke speed climb the wall. Through a touch interface, users can rotate the wall, see it from different angles, and gain insight into the three-dimensionality of the ascent. Other parts of the WebAR let the user explore lead climbing and bouldering. Produced by a five-member team, the WebXR production illustrates the potential to make sports journalism more immersive on the Web (Maese, Rick, Madison Walls, Artur Galocha, Leslie Shapiro, and Ashleigh Joplin, 2021). *The Post* has produced extensive high-quality WebXR content. Its coverage of the destruction of the Amazon rainforest, including WebXR content, won a Polk Award in 2023 (Terrence, 2023; Washington Post Staff, 2023). *The Post* also produced The blast effect: This is how bullets from an AR-15 blow the body apart (Kirkpatrick, Mirza & Canales, 2023). Publication of the WebXR report coincided with the mass school shooting in Nashville, TN, on 27 March 2023, in which the assailant used an AR-15.

Second in this analysis of WebXR journalism is a virtual crime scene. *The New York Times* produced its virtual crime scene via WebAR to allow users to investigate via a first-person touch interface Syria's deadly chemical attack on 7 April 2018. The Syrian military carried out a deadly chemical attack against civilians in Douma, Syria, near Damascus. *Times'* reporters "counted at least 34 victims spread across two floors and in stairways." Via the *Time's* report users can interactively explore a 3D model of the scene of the deadly attack (Browne, Singhvi, Grothjan, Parshina-Kottas, Patanjali, Peyton, Huang, Migliozzi, Koppel, Peng, Buchanan & Wilhelm, 2018). Through the design of the content, those using a smartphone or other mobile device can engage in a fully immersive AR experience; on a desktop or laptop computer, the experience is two-dimensional.

This project is part of *The Times'* research on designing 3D scenes for the Web. One instructive example is an interactive view of Freemans Alley, a street art destination on New York's Lower East Side (Cohrs, Boonyapanachoti, Aneja, Peled, Köerner & Kim, 2021). Through photogrammetry, users of the *Times'* WebXR high-resolution content can travel immersively along the alley exploring it in an omnidirectional format. There is no narrative component to this production but it illustrates the method of creating such WebXR content for journalism.

Third in this analysis is *The Multiplicative Power of Masks*. Produced by Bhatia and Minute Physics (2023) and published by the Berkeley Advanced Media Institute, this is a nonfiction content. Although not published by a news organization, it represents an illustration of WebXR for science and health journalism. Bhatia is a freelance science writer and educator. The content is available via computer or mobile device. It is presented as "An Explorable Essay on How Masks Can End COVID-19." Available in more than a dozen languages, the "explorable essay" illustrates the effective use of a data-driven WebXR experience. Users can interactively explore various transmission routes of COVID-19. Depicting the spread of airborne droplets, an animation shows the user how COVID-19 can spread from a contagious person to a susceptible person. Transmission Route 4 reveals that "When both people wear 50% effective masks, disease transmission drops by 75%" (Bhatia & Minute Physics, 2023). An interactive animation lets the user explore how disease transmission occurs in the entire population. Via a sliding touch bar, the user can adjust the percentage of the population wearing a mask and how changing the level of mask usage affects disease transmission. Drawing on data from the Centers for Disease Control and Prevention and peer-reviewed research on the efficacy of mask usage against COVID-19, the essay illustrates the potential of WebXR journalism. To aid journalism transparency, the article provides interactive details on how the essay was created including a link to the source code available on Github.

Much early XR journalism is no longer accessible due to technological change, and no doubt loss of industry revenues that preclude updating the content for newer platforms. The focus has been increasingly on producing content for the largest audience and VR is still limited to a fragmented market. This is leading to more AR and WebXR production. AR and WebXR are less difficult and expensive to produce than six DoF VR. *Medium* provides a useful set of resources for journalists seeking to produce WebXR content (Singh, 2020). 360 video is similarly simpler and less expensive to produce than six DoF VR, but its storytelling potential is limited. Also, 360 video has become commonplace and has limited promotional novelty or value for news media often seeking to market their products during an era of belt-tightening.

Conclusions

Immersive journalism content is emerging as an important and substantial part of the news industry. VR, AR, and WebXR content are all developing as vital parts of 21st-century journalism, and each may play a key role in the advent of journalistic excellence inside the Metaverse. Based on an analysis of the utilization of the qualities of EM in nine examples of immersive journalism content produced since 2015, there is still substantial untapped potential. Most of the journalism content studied here makes only limited use of the six dimensions of EM. In particular, most VR, AR, and WebXR productions provide only a moderate level of immersion, interactivity, multisensory communication, data/AI, and participatory features, and none use a gaze, gesture, or voice interface. Content that taps into data and uses code and algorithms to present data-driven precision and interactive customization are few in number but demonstrate the powerful potential of such content, especially in the realm of WebXR. Metaverse journalism content that utilizes a full spectrum of EM features may offer the public high-quality news and features and offset the possibility of mis- and disinformation on the immersive Internet. Such journalism content can substantially contribute to the pursuit of truthful journalism inside the Metaverse.

Chapter 8

CONCLUDING REFLECTIONS: ACHIEVING JOURNALISTIC EXCELLENCE IN THE METAVERSE

This chapter offers a vision—a set of recommendations—for the development of journalism of excellence in the Metaverse. Quality journalism is essential for any social system to function effectively and it is just as vital in a virtual realm, such as the Metaverse. Quality here is defined in terms of the truthfulness of journalism that utilizes the affordances of the Metaverse platform, as outlined in the EM framework. The chapter also presents a framework for scholarly inquiry regarding that journalism. This research is vital to the continuing development, improvement, and understanding of Metaverse journalism.

Six Broad Lessons

The vision offered here comprises a set of six lessons drawing upon the analysis in the first seven chapters of the book. First, Metaverse journalism should build on a foundation of three core principles: editorial independence, ethical practice, and full transparency. These principles will fuel journalism in its pursuit of truth in the Metaverse and mitigate potential bias or errors in reporting.

Second, as the shape of Metaverse platforms continues to become more clear, news media need to stake a claim before these realms crystallize and virtual land is out of reach. By establishing a presence early, news media can not only contribute to the shape of the Metaverse but also ensure that its structure, systems, and function embrace independent journalism and freedom of digital press.

Third, reporting in the Metaverse represents both a unique challenge and opportunity for news media. By tapping into the dimensions of immersion, data, interactivity, multisensory communication, AI/data, Natural User Interface (NUI), and first-person perspective, Metaverse journalism can develop a new approach to reporting that delivers DINE, or dynamic immersive news experiences.

Fourth, journalists in the Metaverse need an evolved set of skills to be effective as both reporters and storytellers. Doing this will happen only if human journalists and news avatars can work in a complementary fashion, with human reporters drawing upon AI as a supplementary tool. Used wisely and ethically, AI can prove valuable for quality journalism in the digital vastness of the Metaverse.

Fifth, users of the Metaverse are the public in digital form. For journalism, they are to be engaged in a meaningful fashion and encouraged as partners in the creation of quality journalism. For users, the Metaverse presents exciting possibilities but also poses serious risks. Journalism needs to be the news guardian that complements the technological guardian that ensures a safe technical experience inside a Metaverse platform.

Sixth, journalism has a limited presence on existing Metaverse platforms as of 2023. But the advance of Web3 is fueling the growth of VR, AR, and WebXR journalism content. All three present an important opportunity for the development of Metaverse journalism. News media could build either their own Metaverse platform (e.g., via FrameVR.io) or virtual worlds within an existing platform and populate it with XR journalism content. Producing news content experiences original to the Metaverse is a logical next step in engaging the virtual public.

Three Key Conditions

Applying this set of six lessons can lead to journalistic excellence in the Metaverse. But three key conditions are essential along the way. First, organizations that build and sustain the Metaverse must be committed to creating an environment that nurtures quality journalism. This means freedom of speech and press, operated responsibly, must be a priority within the Metaverse. Although news councils have generally proved ineffective in the real world, there is greater potential for their effectiveness in the Metaverse where a participatory culture will be baked into the virtual world. A Metaverse news council could be made of users, journalists, and others from those interested in the quality of journalism. News councils in the real world include the Minnesota News Council (closed in 2011), the Bolivia News Council, and the British Complaints Commission (Associated Press, 2011). As these have typically operated, a Metaverse news council might serve in a mediating or advisory role to hear and respond to user complaints of bias, inaccuracy, and defamation.

Second, journalism and journalists (human or robotic or a fusion of the two) must be present in a substantial manner in the Metaverse. Unless independent news reporting is produced in a continuous widespread fashion, Metaverse journalism itself will fail to reach a sustainable critical mass and its absence will likely contribute to a downward spiral of the Metaverse as it hurtles toward an almost inevitable end stage of digital implosion due to misinformation, fraud, and abuse. The epic collapse of once-celebrated cryptocurrency exchange FTX in November 2022 provides a case in point (Reiff, 2023). Of course, this is not to imply journalism can prevent fraud or misinformation on the Metaverse or in the real world, but it plays an important role in helping to uncover it for public consideration. News media long have played a vital role in alerting the public to potential problems and Metaverse journalism can continue to perform in this valuable capacity (Lu & Matsa, 2016).

Finally, diverse and inclusive journalists and journalism are essential to the successful production of Metaverse journalistic excellence. Although real-world journalism has often failed in regard to diversity and inclusion, and public trust in news has flagged, Metaverse journalism can and must meet the challenge. If the dual challenge of diversity and inclusion is not met, the value and opportunity of Metaverse journalism excellence will fail and the pursuit of truth will fall short.

If the above three conditions are met, Metaverse journalism content can flourish in both style and substance. Moreover, it is likely that the public will come to trust journalism that is based on the precision of accurate data and highlights transparency. Such transparency in news processes and content can enable the Metaverse public to observe how the news is created and the sources upon which it relies, contribute to that flow of news, see where any bias or inaccuracies may occur and how they can be corrected, and thus help advance journalistic excellence through meaningful public engagement.

Implications for Research on Metaverse Journalism

It would be more than a decade until Stephenson used the term Metaverse to describe a virtual world humans could occupy. But in 1980, the author first entered a computerized virtual world and quickly recognized how powerful an experience it could be and how research about it should be high on the scholarly agenda. ADVENT, short for *Colossal Cave Adventure*, was an early text-based simulation game played on a computer terminal connected to the Internet via a very slow telephone modem (Anderson & Galley, 1995). But regardless of the technical interface and lack of graphics, the game was compelling. The author used playing the game as a reward he held out at night if he completed a certain amount of writing for his dissertation. For the user, the game existed only as a few words on a screen. But it was interactive and psychologically immersive. There was no journalism, but the author did play it in the basement of a journalism school building. The game allowed exploration and action and not only signaled the looming rise of global online video game play. It also pointed the pathway to future scholarly research on a growing area of media engagement. Whether in the form of ethnography or quantitative user metrics, the study of the virtual realm would grow in scope, volume, and importance as the Internet evolved and grew into its increasingly immersive form and as the Metaverse appeared and took shape on the 21st-century digital horizon. The following is a four-part framework or agenda for conducting scholarly research on the Metaverse and the "advent" of journalism within and about it.

Four-Part Framework for Research on Metaverse and Journalism

First on the research agenda is an examination of the **structural and systemic** factors establishing control of the Metaverse, especially the economic and legal considerations in a global context that will frame, enable, or constrain journalism and public engagement. Questions include what is the ownership of the Metaverse platforms and the likely commercial nature of the Metaverse. Metaverse platforms such as *Upland* already sell AR rights to addresses, or actual locations, in the physical world. Real property owners may not realize that someone else might already own their address in the Metaverse, giving them the intellectual property rights to that address online (Roettgers, 2022). Public ownership of Metaverse journalism can heighten the potential for citizen reporting and public participation, areas of deficiency in journalism more generally.

Other questions for research include the following: To what extent are digital enterprises developing the Metaverse committed to ensuring this immersive form of the Internet is guided by an ethical framework? How do policy and regulatory parameters shape the Metaverse? Especially important to journalism and its role in democracy, will access to the Metaverse be free to all or will the Metaverse or Metaverse journalism feature a paywall (i.e., a means to restrict access to content with some sort of required payment; Salwen, Garrison & Driscoll, 2004)? What will be the qualities of that paywall? What will be the fundamental economic underpinnings of the Metaverse? Will the Metaverse evolve into largely a commercial marketplace? Or, to what extent will the Metaverse develop into a culturally rich marketplace of ideas, quality-mediated experiences, and social interaction in a shared community space? Will the Metaverse remain public, or will it become an increasingly private enclave of the wealthy? As the Metaverse will be spatial and may include urban settings, studies of journalism in virtual communities will be important.

Structural and systemic considerations include how journalism organizations are represented in the Metaverse, whether news media generate revenues from that presence, operate for profit or are not for profit, and what legal or regulatory factors may come into play, whether from within the Metaverse or from the external environment (e.g., freedom of speech and press, censorship, libel and privacy laws, copyright). Included as well are self-regulatory considerations including the ethical framework that governs news media action and representation in the Metaverse. If the Metaverse becomes a platform for surveillance capitalism on digital steroids, it will be essential that journalism can act as a check on the growing power of the digital corporate collective. What role a Metaverse news council may play as a means of adjudicating news-related disputes could be an essential research topic.

Research should draw upon the growing field of platform studies. Critical cultural studies of VR and the Metaverse should address issues of the production of culture and the role of commercialism (Egliston & Carter, 2022; Poell et al., 2021). Colonialist concepts such as pioneering and "wild west" often are employed in discourse and journalism regarding VR and the Metaverse (Chia, 2022). How this develops inside the Metaverse is an area for investigation.

Research on surveillance in the Metaverse is another important topic for critical investigation. Andrejevic (2022) has examined how digital enclosures manifest *parallel reality* through technology in the form of a personalized display board at an airport based on surveillance of a user's mobile device. Such surveillance is already in use in Horizon Worlds, where digital signage displays user information for other users to see (Pavlik, 2022).

Second are **design and content** factors that shape and define the UX in the Metaverse including journalism content. Questions to consider include what are the affordances of the design of the Metaverse, especially as they pertain to news. What are the technical standards of the Metaverse and is interoperability required between or across different Metaverse platforms? This will shape access to news across a wide spectrum of the public. Especially important will be design features that ensure equal access to news for all persons regardless of potential disability or economic situation. What are the technical requirements to enter the Metaverse and participate fully within it? This brings substantial implications for bridging the digital divide and may come to define the potential for journalism as an immersive digital good. Is a headset required for a fully immersive experience and how does this affect news engagement and outcomes of that engagement? Will there be publicly available spaces where HMD technologies are available to enter the Metaverse (e.g., public library)? What role do creative immersive content experiences such as cinematic or documentary VR play in the Metaverse? What is the shape and nature of journalistic storytelling within the Metaverse? What role does independent and community-led journalism play as a diverse and reliable source of information about the Metaverse and as a check on the nature, form, and function of the Metaverse that serves all groups, including the marginalized? Will mis- and disinformation flourish within the Metaverse and how does journalism confront it? Such mis- and disinformation falsely presented as journalism has the potential to erode public trust before quality journalism even has a chance to take root in the Metaverse. Open source design can ensure that Metaverse journalism will gain greater public trust and reflect less bias than journalism in other media forms.

Research suggests the Metaverse can play a potentially significant role not only in journalism but also in a wide range of the arenas of news coverage, including education, health, and sport, especially video games and game play. Moreover, as cryptocurrency and its various dimensions such as NFT sales as well as gambling may play a substantial role in the Metaverse, journalists need to make covering the economic character, especially crypto and related arenas, a major area of news coverage.

Third on this research agenda are **usage** considerations. What is the level of public engagement in or usage of an immersive, interactive broadband Internet, or Metaverse? This is the foundation for journalism and sets the upper boundary for how far the reach of journalism in the Metaverse can extend. How does that usage evolve over time and place? How inclusive and open is that usage (including in what languages)? To what extent is the Metaverse a multisensory user experience? What role do algorithms play in shaping the user experience? What level of engagement or control do users have within the Metaverse? All of these dimensions will define the potential form and function of journalism in the Metaverse. What user data are collected and used (e.g., will user privacy rights be respected) will impact not only the economics of journalism and the Metaverse but the likely regulatory framework that limits it. How do more immersive media and news experiences impact users of or participants in the Metaverse? What are the individual rights of users within the Metaverse and how do they frame journalism? To what extent is the Metaverse a curated experience or one that is largely unfiltered or uncensored? To what extent and under what conditions do immersive news and other media experiences in the Metaverse generate user presence and other outcomes (e.g., psychological, emotional, learning)? Is accessibility a foundational principle that guides the Metaverse and journalism within it? Accessibility should be the starting place for inclusive Metaverse journalism.

User research is particularly salient and should build on an extensive existing body of UX studies. Methods, theories, and effects of VR experiences are all salient to the design and impact of Metaverse journalism. Because VR, and ultimately the Metaverse, can seem real and at times intensely so, it is particularly incumbent on those designing experiences, including journalism, to be fully aware of the possible outcomes, both positive and negative, before publishing content.

A variety of methodological approaches are warranted, including experimental design, observational and ethnographic studies, survey research, and content analysis. Studies that monitor users' eye, body, or facial motion, physical position, or distance from objects and encounters with objects may be useful. Existing research in this arena is intended to improve the function of VR (e.g., W. Kim, 2022) or assess human response to VR (e.g., Pastel, 2022). Some studies have monitored participants' physiological response as measured in the brains or cardiovascular system when experiencing VR (e.g., Lemmens, Simon & Sumter, 2022). Studies have employed multi-method approaches in such research such as Lemmens' "Fear and Loathing in VR." This work used subjective measures, surveys of user views and emotions, and objective measures such as heart rate. Using multi-method approaches can yield more reliable results but are also more expensive to design and conduct.

News media will need to account for potential cyber or VR sickness (Seok, Kim, Son & Kim, 2022). VR sickness may be a form of motion sickness, per Sensory Conflict Theory, or sensory mismatch theory (Koch, Cascorbi, Westhofen, Dafotakis, Klapa & Kuhtz-Buschbeck, 2018). Mismatched senses associated with body movement may play an important role (Brown et al., 2022). This research may present important implications for Metaverse journalism especially for VR design, HMD usage, and its impact on the length of a news experience. Cyber sickness can be assessed via EEG measurement or self-report. Subjective standardized questionnaires (like the Motion Sickness Questionnaire) and measured actions (walking, pegboard tests, performance in a test) also are relevant. Experiments that compare the performance of VR systems are relevant (e.g., Ahn, Nowak & Bailenson, 2022).

It is also essential that ethics guide research on human subjects, including that conducted by industry. Academic investigators are required to have their research methods independently reviewed for ethical treatment of subjects before conducting a study. News industry (in fact, all industries in the Metaverse) studies should employ the same model of independent review, even if not mandated by law.

Assessing the nature and impact of nonverbal communication such as a gesture, a facial expression, or a nod, with regard to variables relevant to news such as believability, trust, and perceived authenticity are relevant to journalism in the Metaverse (Huang & Jung, 2022; Brodsky, 2021). Studies of the effectiveness of learning via VR (e.g., Lee & Hwang, 2022) are relevant to journalism. Studies reveal that human cognition, including creativity, is embedded within the body and its sensorimotor processes (Gibbs, 2005). Thus, embodiment is fundamental to the UX in the Metaverse and research suggests it is vitally important to the design of Metaverse journalism.

Defining key terms such as presence is important to this research and the potential impact of Metaverse journalism. Skarbez, Brooks, and Whitton (2017) suggest different views based on philosophical or psychological approaches. Among the most common is presence as the feeling of being there or in a space or location, though it can also take other forms including the "illusion of nonmediation" (Murphy & Skarbez, 2022: 43). Moreover, the intensity of the user's sense of presence can have a profound impact on their response to a virtual experience and it may be a curvilinear one in which past a certain point a positive experience can become a negative or even harmful one. These frames have important implications for designing immersive journalism content as the type of presence the user experiences may influence outcomes such as empathy or understanding. Studying the link between presence and the illusion of plausibility of a VR experience is especially pertinent to Metaverse journalism. The appropriate level of abstraction of immersive content is worth studying in a news context (Slater, Banakou, Beacco, Gallego, Macia-Varela & Oliva, 2022).

How users respond to the perspective or identity of their avatar pertains to journalism. Research shows that if virtual skin color differs from a user's real-world skin color, it may affect empathy and has important implications for how Metaverse journalism covers social justice issues (Ambron, Goldstein, Miller, Hamilton & Branch, 2022). Avatar realism can lead to the uncanny valley (Mori et al., 2012). The role of more realistic avatar faces merits further investigation in this context and may suggest important design considerations for Metaverse journalism (Ovide, 2022).

The Proteus effect is worth studying in the Metaverse. Research shows that user attitudes and behaviors in the real world tend to align with those of their avatars (Kyrlitsias & Michael-Grigoriou, 2018). How this extends to the Metaverse could provide important insights for journalism. The doorway effect also should be studied. Research shows that when a human passes through a doorway, their mind tends to reset and they often forget what motivated them to move from one room to another. Recent studies show that the same effect can occur in VR. How it might pertain to the Metaverse could be useful in how journalists design scene reconstructions and user experiences in them.

VR and sports in the Metaverse may relate to addiction and particularly to gambling addiction with the rise of online and mobile gaming and the advent of gambling in the Metaverse. How does journalism navigate this ethically fraught domain that brings coverage, users, and revenues into potential conflict of interest? Addiction in the Metaverse may be a particular problem among youth (Kaimara, Oikonomou & Deliyannis, 2022). Moreover, children may have a limited capacity to distinguish between make-believe and reality (Barreda-Ángeles & Hartmann, 2022). This could raise important ethical considerations in designing Metaverse journalism experiences and how they are labeled.

Immersion is a key concept and has been defined and studied as the notion of sensory envelopment (Norton, Sauer & Gerhard, 2022). Yet, different platforms immerse differently or in varying degrees to alternative senses. How this applies to Metaverse journalism in designing multisensory news experiences merits further research.

Ethnographic or observational approaches to study content generated within the Metaverse are particularly relevant to journalism. Such approaches could reveal important lessons about creating productions such as *We Met in VR*. Understanding the user response would point to how reporting might move forward more effectively in the Metaverse.

Fourth on the research agenda for Metaverse journalism are **relational** considerations. What is the nature of social interaction within the Metaverse and the news media? As the public has increasingly demonstrated a preference for news delivered via social media, it is similarly likely that such social delivery forms are likely to play a significant role in the Metaverse. What is the relationship between internal Metaverse experiences and social media platforms that connect to or interface with the wider real world? Do they become a bridge for news flow? In a sign of things to come, Meta already has begun adding 3D avatars to Instagram Stories (Lyons, 2022). Studies of avatars in the context of journalism will be valuable (Coleman, 2011). The relationship between Metaverse journalism content and gender identity merits study. How will journalism report on stories that revolve around the adverse side of the Metaverse? How will Metaverse users be shielded from unwanted social interaction and protected from sexual assault or harassment? Meta's platform provides a virtual "personal boundary" shield, but will it be enough to ensure a healthy and robust experience for all (Sims, 2022)? Will automated AI-driven experiences and avatars be enabled and what role will they play in the Metaverse, especially with regard to journalism? Will fake news, including intentional disinformation campaigns aligned with state players, emerge as a significant part of the Metaverse landscape and will users be able to sort the real from the fake in a completely synthetic virtual realm and avoid sharing misinformation?

With this four-part agenda in mind, journalism and media scholars can help to inform the public, industry, and policymakers about the nature and impact of the Metaverse. Moreover, such scholarship can contribute essential insight to help advance understanding of the evolving nature of the Metaverse and the role that journalism will play in and about it. Journalism, media, and communication scholars also can shape industry practice and public policy to guide and help ensure the 21st-century development of a Metaverse, particularly with regard to journalism, that is (1) socially beneficial, (2) rich in quality immersive news experiences, and (3) nurtures imaginative and inclusive content creators. The alternative is likely to lead to a dystopian Metaverse that approximates the fictional *Ready Player One* Hollywood blockbuster (Cline, 2011, 2018) in which journalism plays virtually no role. Lost may be the potential to create the socially beneficial and individually empowering vision of VR developer Jaron Lanier. The fact the Metaverse may fall under Clarke's third law (i.e., any sufficiently advanced technology is indistinguishable from magic) should not deter journalists, news media, and any other interested parties from making the right choices that will shape the development of the Metaverse to the betterment of humankind and journalistic excellence in a post-COVID-19 pandemic world.

Research on the Metaverse and its consequences is still embryonic and as such important gaps in the research literature exist. The theoretical principles and mechanisms that shape and enable the efficacy of Metaverse-based experiences need further investigation. Of particular relevance for journalism in the Metaverse are questions of what content forms most impact the UX and how, the nature of affordances in Metaverse, and how these affordances intersect with news engagement.

Theories of the Metaverse and Metaverse journalism are developing based on key concepts such as presence, embodiment, and plausibility, based on mostly psychological and physiological perspectives. Can Metaverse journalism become an "empathy machine" and under what conditions or is this hype (e.g., Raz, 2022)? Interrogating the binary notion that divides reality and virtual merits consideration. Whether this is a false or misleading binary is made abundantly clear in the notion of "mixed reality." A binary conceptualization limits possibilities for journalism in the Metaverse.

Research on how Metaverse experiences build meaning is especially germane to journalism. Old rules and models may no longer pertain. For instance, the five Ws and the inverted pyramid of news story construction may be largely obsolete in Metaverse journalism. Instead, how to use sound, imagery, graphical design, and haptics to direct the user's attention in an interactive and immersive virtual world may shape Metaverse journalism and warrant research. Kinetic syntax for haptic communication needs research and development and especially so if Metaverse journalism is to use it effectively (D'Armenio, 2022). A kinetic reconceptualization for VR and video news games merits consideration. In *New Techno Humanities*, D'Armenio observes (2022): "*Tetris*, despite being an abstract video game, has a semantic component [...] it is effective precisely because it is based on a universal dynamic: the struggle against chaos. It is a semantic parable expressed by a dynamic that is both visual and kinetic" (p. 7). How journalism transforms in a virtual world where news users are physically active and participatory in their news engagement is a very different model of journalism than exists and gives meaning in the physical world.

The news industry has a long history of collaboration with scholars. In the 1940s, the Lazarsfeld–Stanton Program Analyzer was used in the study of radio audiences and in the new medium of the day, television (Levy, 1982). The Lazarsfeld–Stanton Program Analyzer was a product of collaboration between Paul Lazarsfeld, Columbia University Professor from 1949 to 1969, and Frank Stanton, President of CBS from 1946 to 1971 (Columbia University, 2004; Noble, 2006). Their research tool contributed to the development of the electronic media and ultimately laid a foundation for new forms of audiovisual journalism (Rogers, 1994).

Journalism leaders could advance a model of research collaboration in the Metaverse. Collaborative research could help the development of news experiences that are designed to maximize user engagement in the news. Scholars could have access to more resources to support their research and help develop a program of systematic inquiry on the development of news experiences in the Metaverse.

Conclusions: The Way Forward for Metaverse Journalism

The Metaverse today is an opportunity for journalism leaders to get out ahead of the transformation of media. It is not without risk, but waiting until others take that risk and determine what form of news works in a virtual world could leave today's journalists and news media out of the fledgling new news industry, or it might never even develop.

Finding the courage to embark on a new journalistic course might draw upon the innovative spirit of journalists past. Legendary journalist Lowell Thomas, famous for his reporting of Lawrence of Arabia, in the 1950s, developed "Cinerama," the forerunner to today's VR journalism (Stephens, 2016; *The Okaloosa* News-Journal, 1956). Edward R. Murrow helped invent both radio and television news in the first half and middle of the 20th century.

In the late 19th century, journalist and newspaper publisher Ida B. Wells persevered against White Supremacists who burned down her paper's printing press and continued her fight to document the history of lynching in America. Just 22 years old at the time, Benjamin Day employed advances in printing technology in 1833 to found the *New York Sun*, the first penny press newspaper, paving the way for the rise of mass media and the advent of a new news business model based on the sale of advertising in his newspaper. Even before that, Elias Boudinot founded the *Cherokee Phoenix* in 1828 as America's first Native American newspaper. These are just a few examples of the spirit of innovation that has characterized journalism of the past two centuries. A new generation of news innovators may invent the immersive journalism form that can succeed and provide a pathway to the truth in the Metaverse.

REFERENCES

.cult (24 Feb. 2022). 10 metaverse jobs that will exist by 2030. https://thenextweb.com/news/10-metaverse-jobs-that-will-exist-by-2030.

ABCNews (17 July 2021). Anthony Bourdain Film Under Fire. https://abcnews.go.com/GMA/Culture/video/director-fire-ai-recreate-anthony-bourdains-voice-78880088.

Ahn, S. J. G., Bailenson, J. N., & Park, D. (2014). Short-and long-term effects of embodied experiences in immersive virtual environments on environmental locus of control and behavior. *Computers in Human Behavior*, 39, 235–245.

Ahn, Sun Joo Grace, Kristine L. Nowak, & Jeremy N. Bailenson (2022). Unintended consequences of spatial presence on learning in virtual reality. *Computers & Education*. doi: 10.1016/j.compedu.2022.104532.

Akour, I. A., Al-Maroof, R. S., Alfaisal, R., & Salloum, S. A. (2022). A conceptual framework for determining metaverse adoption in higher institutions of gulf area. *Computers and Education: Artificial Intelligence*, 3, 100052. doi: 10.1016/j.caeai.2022.100052.

Ackerman, Daniel (10 March 2021). Using artificial intelligence to generate 3D holograms in real-time. https://news.mit.edu/2021/3d-holograms-vr-0310.

Al-Heeti, Abrar (17 July 2021). "Use of AI to copy Anthony Bourdain's voice for documentary sparks criticism." https://www.cnet.com/news/anthony-bourdain-documentary-used-ai-to-re-create-his-voice-after-he-died/.

Al-Saqaf, Walid, Malin Picha Edwardsson (2019). Could blockchain save journalism? Chapter in Massimo Ragnedda, Giuseppe Destefani (editors). Blockchain and Web 3.0. Routledge. ISBN 9780429029530.

All News (7 Fed. 2023). S. Korea opens metaverse platform for Korean-language learning. https://en.yna.co.kr/view/AEN20230207005900315.

Allan, Sean (5 September 2021). The Three Tiers of AI: Automating tomorrow with AGI, ASI & ANI. Aware. https://www.aware.co.th/three-tiers-ai-automating-tomorrow-agi-asi-ani/.

Allison, Ian (2 November 2022). "Divisions in Sam Bankman-Fried's Crypto Empire Blur on His Trading Titan Alameda's Balance Sheet".

ACM Awards (2019). http://amturing.acm.org/award_winners/sutherland_3467412.cfm.

Ambron, Elisabetta, Shayna Goldstein, Alexander Miller, Roy H. Hamilton, & Branch (2022). From my skin to your skin: Virtual image of a hand of different skin color influences movement duration of the real hand in Black and White individuals and influences racial bias. *Frontiers in Virtual Reality*, (3). doi: 10.3389/frvir.2022.884189.

Amenabar, Teddy (22 February 22). Sony into metaverse as gaming platform.

Anderson, Tim & Stu Galley (1995). "The History of Zork".

Andrejevic, M. (2022). Meta-Surveillance in the Digital Enclosure. *Surveillance & Society*, 20(4), 390–396.

Apple AI-narrated audio book, 2023. https://itunes.apple.com/WebObjects/MZStore. woa/wa/viewAudiobook?id=1640812516.

Archer, Dan and Katharina Finger (15 March 2018). "Walking in another's virtual shoes: Do 360-degree video news stories generate empathy in viewers?" Columbia Journalism Review. https://www.cjr.org/tow_center_reports/virtual-reality-news-empathy.php.

Aristotle, *The Politics [c. 330 BCE]*, Stephen Everson (ed.), Cambridge: Cambridge University Press, 1988.

Asimov, Isaac (1950). I, Robot.

Associated Press (28 January 2011). "Minnesota News Council Closing After 40 Years." https://www.cbsnews.com/minnesota/news/minnesota-news-council-closing-after-40-years/.

Au, Wagner James (2008). *The Making of Second Life*. New York: Harper Business.

Bahuguna, Anmol (2022). Intellectual Property and Metaverse. https://www.ijsr.net/archive/v11i9/SR22905165231.pdf. International Journal of Science and Research (IJSR) ISSN: 2319-7064 SJIF: 7.942.

Bailenson, Jeremy. Experience on Demand: What Virtual Reality Is, How It Works, and What It Can Do. W.W. Norton & Company, 2108. 290 pp.

Bajarin, Tim (3 Jan. 2023). Is Meta Pivoting To AR? https://www.forbes.com/sites/timbajarin/2023/01/03/is-meta-pivoting-to-ar/.

Ball, Matthew (2022). *The Metaverse: And How It Will Revolutionize Everything*. Liveright Publishing.

Banakou, D., Hanumanthu, P. D., & Slater, M. (2016). Virtual embodiment of white people in a black virtual body leads to a sustained reduction in their implicit racial bias. *Frontiers in human neuroscience*, 10, 601. Berners-Lee, Tim (2022). "History of the Web." https://webfoundation.org/about/vision/history-of-the-web/.

Barr, Kyle (22 September 2022). TikTok company's new VR headset competes with meta on price and privacy. Gizmodo.com https://gizmodo.com/tiktok-vr-bytedance-pico-headset-meta-quest-2-pro-1849570315.

Barreda-Ángeles, M., & Hartmann, T. (2022). Hooked on the metaverse? Exploring the prevalence of addiction to virtual reality applications. *Frontiers in Virtual Reality*, 3. https://doi.org/10.3389/frvir.2022.1031697.

Barthel, Michael (JULY 27, 2021). 6 key takeaways about the state of the news media in 2020. Pew Research Center. https://www.pewresearch.org/fact-tank/2021/07/27/6-key-takeaways-about-the-state-of-the-news-media-in-2020.

Baum, Tyler (18 March. 2022). I made $1.4 million from selling virtual items inside Roblox metaverse. NYPost.com.

BBC News (2015). Calais Migrants: What's it like in the "Jungle"? (360 video) BBC News. https://www.youtube.com/watch?v=9McdcF3CglE.

—— (7 Feb. 2018). BBC launches augmented reality app for *Civilisations*. http://www.bbc.com/news/technology-42966371.

Beech, Hannah (18 July 2018). Step Inside the Thai Cave in Augmented Reality. https://www.nytimes.com/interactive/2018/07/21/world/asia/thai-cave-rescue-ar-ul.html.

Belanger, Ashley (6 Feb. 2023). Twitter suspended 400K for child abuse content but only reported 8K to police. https://arstechnica.com/tech-policy/2023/02/twitter-only-reported-2-of-accounts-suspended-for-child-abuse-content-org-says/.

Bellamy, Claretta (Nov. 28, 2022). Morehouse College class will teach Black history in the metaverse. https://www.nbcnews.com/news/nbcblk/morehouse-college-class-will-teach-black-history-metaverse-rcna57159.

Bensinger, Ken (10 Feb. 2023, DeSantis, Aiming at a Favorite Foil, Wants to Roll Back Press Freedom. NYTimes.com https://www.nytimes.com/2023/02/10/us/politics/ron-desantis-news-media.html.

Best, Shivali (22 March 2022). Japanese start-up develops armband that mimics PAIN in the metaverse. https://www.dailymail.co.uk/sciencetech/article-10639295/Japanese-start-develops-armband-mimics-PAIN-metaverse.html.

Bevensee, Emmi and Max Aliapoulios (Oct. 26, 2020). Introducing the U.S. Election Twitter Network Graph Tool. https://foundation.mozilla.org/en/blog/fellow-introducing-us-election-twitter-network-graph-tool/.

Bhardwaj, Pranav (16 Feb. 2023). The Top 10 Dangers of Using Metaverse. makeuseof.com https://www.makeuseof.com/dangers-of-metaverse/.

Bhatia, Aatish & Minute Physics (2023). The Multiplicative Power of Masks: An Explorable Essay on How Masks Can End COVID-19. Berkeley Advanced Media Institute. https://aatishb.com/maskmath/.

Bhuiyan, Md Momen, Hayden Whitley, Michael Horning, Sang Won Lee, and Tanushree Mitra (2021). Designing Transparency Cues in Online News Platforms to Promote Trust: Journalists' & Consumers' Perspectives. *Proc. ACM Hum.-Comput. Interact.* 5, CSCW2, Article 395 (October 2021), 31 pages. https://doi.org/10.1145/3479539.

Big Rock Creative (2023). Breonna's Garden https://bigrockxr.com/breonnas-garden/.

Bilton, Ricardo (30 November 2016). Reuters built its own algorithmic prediction tool to help it spot (and verify) breaking news on Twitter. NiemanLab.org. https://www.niemanlab.org/2016/11/reuters-built-its-own-algorithmic-prediction-tool-to-help-it-spot-and-verify-breaking-news-on-twitter/.

Biocca, F., & Delaney, B. (1995). Immersive virtual reality technology. *Communication in the age of virtual reality*, 15, 32.

Biocca, F., Kim, J., & Choi, Y. (2001). Visual touch in virtual environments: An exploratory study of presence, multimodal interfaces, and cross-modal sensory illusions. *Presence: Teleoperators & Virtual Environments*, 10(3), 247–265.

Black Enterprise Editors (March 10, 2022). Breonna's Garden Launches Intimate VR Experience at SXSW. https://www.blackenterprise.com/breonnas-garden-launches-intimate-vr-experience-at-sxsw/.

Black Metaverse (2023). https://www.blackmetaverse.io/.

Blake, Tom (18 November 2022). Q.ai Review—An AI-Powered Investing App From Forbes. https://investorjunkie.com/investing/q-ai-review/.

Boonyapanachoti, Mint, Jon Cohrs, Minkyoung Kim, Niko Koppel, Mark McKeague, Lana Z Porter, Guilherme Rambelli, Ben Wilhelms (27 July 2020). Reconstructing Journalistic Scenes in 3D. https://rd.nytimes.com/projects/reconstructing-journalistic-scenes-in-3d.

Boorstin, Daniel J. (1962). The Image. ISBN 978-0679741800.

Brian Bowers (2001). Sir Charles Wheatstone FRS: 1802–1875 (2nd ed.). IET. pp. 207–208. ISBN 978-0-85296-103-2.

Bowman, Emma (9 Jan. 2023). A college student created an app that can tell whether AI wrote an essay. https://www.npr.org/2023/01/09/1147549845/gptzero-ai-chatgpt-edward-tian-plagiarism.

Bracy, Jedidiah (20 Jan. 2023). A look at what to expect for privacy in 2023 with Omer Tene. IAPP. https://iapp.org/news/a/a-look-at-what-to-expect-for-privacy-in-2023-with-omer-tene/.

Branch, John (5 Feb. 2018). Augmented Reality: Four of the Best Olympians, as You've Never Seen Them. https://www.nytimes.com/interactive/2018/02/05/sports/olympics/ar-augmented-reality-olympic-athletes-ul.html.

Bregman, Dion M., Jason E. Gettleman, Douglas J. Crisman and John L. Hemmer (August 2022). Metaverse: A Jumpstart Guide to Intellectual Property, Antitrust, and International Considerations. https://www.morganlewis.com/pubs/2022/08/metaverse-a-jumpstart-guide-to-intellectual-property-antitrust-and-international-considerations.

Brennen, Bonnie S. and Erika dela Cerna (1 August 2010). "Journalism in Second Life." Journalism Studies, 11:4, 546–554, doi: 10.1080/14616701003638418; E-Publications@Marquette https://epublications.marquette.edu/cgi/viewcontent.cgi?referer=&httpsredir=1&article=1065&context=comm_fac.

Brennan, Pat (9 Feb. 2023). 2M1207 b—First image of an exoplanet. NASA. https://exoplanets.nasa.gov/resources/300/2m1207-b-first-image-of-an-exoplanet/.

Brobowsky, Meghan (12 Feb. 2023). TikTok's Parent Takes on Meta in Battle for Virtual-Reality Market Pico, owned by ByteDance, has increased shipments of VR headsets as it attempts to compete with Meta's Quest 2. https://www.wsj.com/articles/tiktoks-parent-takes-on-meta-in-battle-for-virtual-reality-market-dd4abdb6.

Brodsky, Sascha (15 March 2021). Facial Expressions Could Make VR More Accessible and Immersive. https://www.lifewire.com/how-face-tracking-could-make-vr-better-5116169.

—— (28 January 2022). "You May Soon Be Able to Get into the Metaverse Without a Headset." https://www.lifewire.com/you-may-soon-be-able-to-get-into-the-Metaverse-without-a-headset-5217382.

—— (4 February 2022). "Making the Metaverse Accessible Is Better for Everyone." https://www.lifewire.com/making-the-Metaverse-accessible-is-better-for-everyone-5218160.

—— (March 9, 2022). Mixed Reality Could Turn Robots into Extensions of You: Remote control via a headset. https://www.lifewire.com/mixed-reality-could-turn-robots-into-extensions-of-you-5221643.

Browne, Malachy, Anjali Singhvi, Evan Grothjan, Yuliya Parshina-Kottas, Karthik Patanjali, Miles Peyton, Jon Huang, Blacki Migliozzi, Niko Koppel, Jessica Peng, Larry Buchanan and Ben Wilhelm (24 June 2018). How We Created a Virtual Crime Scene to Investigate Syria's Chemical Attack. https://www.nytimes.com/interactive/2018/06/24/world/middleeast/douma-syria-chemical-attack-augmented-reality-ar-ul.html.

Brown, Lane (27 Dec. 2022). Soon you'll be able to make your own feature length movie with AI https://www.msn.com/en-us/movies/news/soon-you-ll-be-able-to-make-your-own-feature-length-movie-movie-with-ai/ar-AA15I4Yu.

Buchanan, Larry, James Glanz, Evan Grothjan, K.K. Rebecca Lai, Allison Mccann, Karthik Patanjali, Yuliya Parshina-Kottas, Jeremy White and Graham Roberts (17 April 2019). Why Notre-Dame Was a Tinderbox. https://www.nytimes.com/interactive/2019/04/17/world/europe/notre-dame-cathedral-fire-spread.html.

Butler, Sydney (3 April 2022). Persistent AR Explained. https://www.howtogeek.com/788486/persistent-ar-explained-why-its-the-key-to-the-metaverse/.

Byrne, R.W. (2013). A Nod to Ned Ludd. The Baffler, 23, 120–128.

Byrnes, Olivia, Wendy La, Hu Wang, Congbo Ma, Minhui Xue, Qi Wu (2021). Data Hiding with Deep Learning: A Survey Unifying Digital Watermarking and Steganography. https://arxiv.org/pdf/2107.09287v1.pdf; https://deepai.org/publication/data-hiding-with-deep-learning-a-survey-unifying-digital-watermarking-and-steganography.

Caddy, Becca and Gerald Lynch (8 April 2022). This haptic glove lets you feel the virtual reality metaverse. https://www.techradar.com/news/this-haptic-glove-lets-you-feel-the-virtual-reality-metaverse, Mike (November 10, 2019). AR and VR will make spatial journalism the future of reporting. https://venturebeat.com/business/ar-and-vr-will-make-spatial-journalism-the-future-of-reporting/.

Cadoes, Olivia, Wendy La, Hu Wang, Congbo Ma, Minhui Xue, Qi Wu (2021). Data Hiding with Deep Learning: A Survey Unifying Digital Watermarking and Steganography. https://arxiv.org/pdf/2107.09287v1.pdf;https://deepai.org/publication/data-hiding-with-deep-learning-a-survey-unifying-digital-watermarking-and-steganography.

Cadoux, Mike (10 November 2019). Augmented reality and virtual reality will make spatial journalism the future of reporting. *Venture Beat.* Retrieved from https://venturebeat.com/2019/11/10/ar-and-vr-will-make-spatial-journalism-the-future-of-reporting.

Campa, Emilio (9 April 2022). How the metaverse could disrupt the in-car experience. https://venturebeat.com/2022/04/09/how-the-metaverse-could-disrupt-the-in-car-experience/.

Campoamor, Danielle (March 24, 2022). New report reveals the dangers of virtual reality for young children. https://www.today.com/parents/parents/dangers-virtual-reality-young-children-rcna21278.

Canales, Katie (28 Feb. 2022). Mark Zuckerberg says it's 'reasonable' that the metaverse isn't a place but a point in time. Business Insider. https://www.businessinsider.com/mark-zuckerberg-reasonable-construct-metaverse-time-not-place-podcast-interview-2022-2.

Capoot, Ashley (14 Dec. 2022). Twitter suspends account dedicated to tracking Elon Musk's private jet. https://www.cnbc.com/2022/12/14/twitter-suspends-elonjet-account-that-tracks-elon-musks-private-jet-.html.

Carey, J. (1989). *Communication as Culture.* Routledge, New York and London.

Carey, John, M. C. J. Elton (2010). When Media are New: Understanding the Dynamics of New Media Adoption and Use. Digital Culture Books/University of Michigan Press and the University of Michigan Library. 0472070851, 9780472070855.

Carlton, Bobby (31 March 2022). Could VR Sports Be The Future Of The Metaverse? https://vrscout.com/news/could-vr-sports-be-the-future-of-the-metaverse/.

Carter, Bryan (2013). Expression through Machinima: A Digital Africana Studies Pedagogical Case Study. *Fire!!!,* Vol. 2, No. 1 (February 2013), (100% effort) pp. 86–104.

Carter, Bryan, Judith Molka-Danielsen and Alastair Creelman (2009b). Empathy in Virtual Learning Environments. *International Journal Networking and Virtual Organisations.* Vol. 6, No. 2 (2009), (33% effort) pp. 123–139.

Carter, Bryan, Judith Molka-Danielsen, David Richardson and Bjorn Jaeger (2009a). Teaching and Learning Affectively Within a Virtual Campus. *Int. J. of Networking and Virtual Organisations.* Vol 5, No. 5 (2009), (25% effort) pp. 476–498.

CBS News (10 June 2021). Artists use artificial intelligence to bring their creations to life. https://www.cbsnews.com/video/artists-use-artificial-intelligence-to-bring-their-creations-to-life/.

CEU Universities (15 October 2021). "CEU Universities already have their own metaverse." https://www.ceuuniversities.com/en/ceu-universities-already-have-their-own-metaverse/.

Chen, Brian X. (18 Dec. 2019). What We Learned About the Technology That Times Journalists Use; https://www.nytimes.com/2019/12/18/technology/personaltech/technology-times-journalists-use.html?referringSource=articleShare.

Chia, A. (2022). The metaverse, but not the way you think: game engines and automation beyond game development. *Critical Studies in Media Communication*, 39(3), 191–200.

Cho, S. (2022). Immersive Virtual Reality: Criminal Law Must Keep Up with Technology. *Harvard Undergraduate Law Journal.* https://hulr.org/spring-2022/sexual-assault-in-immersive-vr.

Christensen, Clayton M., Michael E. Raynor, & Rory McDonald (December 2015). What is disruptive innovation? Harvard Business Review. Retrieved from https://hbr.org/2015/12/what-is-disruptive-innovation.

Chryststomou, George (20 October 2021). Encanto: Everything You Need To Know About Disney's Latest Animated Feature." https://screenrant.com/encanto-everything-need-know-disney-latest-animated-feature/.

Clark, Mitchell (3 May 2021). Fortnite made more than $9 billion in revenue in its first two years. https://www.theverge.com/2021/5/3/22417447/fortnite-revenue-9-billion-epic-games-apple-antitrust-case.

Clarke, Arthur C. (1961). Profiles of The Future.

Clement, J. (17 November 2022). "Video game industry—Statistics & Facts." Statista https://www.statista.com/topics/868/video-games/.

—— (2 Aug. 2022). Number of monthly active players of Minecraft worldwide as of August 2021. https://www.statista.com/statistics/680139/minecraft-active-players-worldwide/.

—— (2023). Global active users of Meta Horizon Worlds VR platform 2022. https://www.statista.com/statistics/1362347/meta-horizon-worlds-users/.

Cline, Ernest (2011; 2018). *Ready Player One.* New York: Ballantine Books.

Cohrs, Jonathan, Mint Boonyapanachoti, Sukanya Aneja, Avner Peled, Willa Köerner, Minkyoung Kim (10 Sept. 2021). Delivering 3D Scenes to the Web. https://rd.nytimes.com/projects/delivering-3d-scenes-to-the-web).

Coinquora (27 August 2021). CryptoPunks NFTs Record $900 Million Month Sales. https://www.investing.com/news/cryptocurrency-news/cryptopunks-nfts-record-900-million-month-sales-2601350.

Cole, Jarrrard (4 Nov. 2015). "Virtual Reality: Behind the Scenes with a Ballerina at Lincoln Center." WSJ. http://www.wsj.com/articles/behind-the-scenes-with-a-ballerina-at-lincoln-center-1446646806.

Coleman, Beth (2011). *Hello avatar: rise of the networked generation.* The MIT Press.

Collins, Keith (17 Feb. 2023). How to Spot Robots in a World of A.I.-Generated Text. NYTimes. https://www.nytimes.com/interactive/2023/02/17/business/ai-text-detection.html.

Columbia Daily Spectator (2 December 1998). Volume CXXII, Number 130. https://spectatorarchive.library.columbia.edu/cgi-bin/columbia?a=d&d=cs19981202-01.2.24&e=-------en-20--1--txt-txIN-------.

Columbia University (2004). Paul Lazarsfeld. Columbia 250. https://c250.columbia.edu/c250_celebrates/your_columbians/paul_lazarsfeld.html.

Cook, Allan V. (2018). Digital Reality and the Revival of Journalism. https://www.wired.com/wiredinsider/2018/08/digital-reality-and-the-revival-of-journalism/.

Crewe, David (31 January 2022). "This 8K 3D Stereoscopic Monitor is a Wild Look at the Future of Displays." https://petapixel.com/2022/01/31/this-3d-stereoscopic-monitor-is-a-wild-look-at-the-future-of-displays/.

Crossplay: An armistice for the console wars. Turtle Beach Blog. (2019). https://blog.turtlebeach.com/crossplay-armistice-for-console-wars/.

Culliford, Elizabeth and Nivedita Balu (25 October 2021). "Facebook invests billions in Metaverse efforts as ad business slows." Reuters. https://www.reuters.com/technology/facebook-revenue-misses-estimates-apples-privacy-rules-bite-2021-10-25/.

Cultural Services French Embassy in the United States (2020). "Tribeca Cinema360 Online: "Saturnism", an Interpretation of Goya's Painting." https://www.digitalbodies.net/tribeca-film-festival-cinema360-moves-online-this-year/.

D'armenio, E. (2022). The mediatic dimension of images: visual semiotics faced with Gerhard Richter's artwork. *Visual Communication*, 0(0). https://doi.org/10.1177/14703572221080529.

Daily 360, The (n.d.). The New York Times. https://bit.ly/3Cn3Xs4.

Dakka, Wisam and Luis Gravano (2007). Efficient Summarization-Aware Search for Online News Articles. http://www1.cs.columbia.edu/~gravano/Papers/2007/jcdl07.pdf.

Dalton, Jeremy & Louise Liu, 2019. PwC "Seeing Is believing" report. https://www.pwc.com/gx/en/industries/technology/publications/economic-impact-of-vr-ar.html.

Dandurand, Darragh (March 28, 2022). Arizona State Launching New VR/AR Classes, Nonny de la Peña to Helm. https://vrscout.com/news/arizona-state-launching-new-vr-ar-classes-nonny-de-la-pena-to-helm/.

Danise, Amy (22 March 2022). Council Post: Five Industries That Will Be Transformed By The Metaverse. https://www.forbes.com/sites/forbestechcouncil/2022/03/22/five-industries-that-will-be-transformed-by-the-metaverse/.

Davies, Chris (2 February 2022). "Meta stock mauled as Facebook reports billions in Metaverse losses." https://www.slashgear.com/meta-stock-mauled-as-facebook-reports-billions-in-Metaverse-losses-02710239/.

Davies, Pascale (17 December 2021). "Sexual assault has already started in Meta's Horizon Worlds Metaverse." https://www.euronews.com/next/2021/12/17/sexual-assault-has-already-started-in-meta-s-horizon-worlds-metaverse.

Davis, F. D.; Bagozzi, R. P.; Warshaw, P. R. (1989), "User acceptance of computer technology: A comparison of two theoretical models", *Management Science*, 35 (8): 982–1003, doi: 10.1287/mnsc.35.8.982, S2CID 14580473.

de Gaynesford, Robert Maximillian (2006). *I: The Meaning of the First Person Term*. Oxford: Oxford University Press.

De Vynck, Gerrit, Rachel Lerman and Nitasha Tiku (February 16, 2023). Microsoft's AI chatbot is going off the rails. WashingtonPost.com. https://www.washingtonpost.com/technology/2023/02/16/microsoft-bing-ai-chatbot-sydney/.

de la Peña, Nonny (2012). Hunger in Los Angeles. https://docubase.mit.edu/project/hunger-in-los-angeles/.

—— (2010). "Immersive Journalism: Immersive Virtual Reality for the First-Person Experience of News," *Presence: Teleoperators & Virtual Environments*, no. 4, 291–301.

Denton, Jack (21 February 2022). Alibaba Stock and Tencent Are Tumbling. This Time, Blame the Metaverse. https://www.barrons.com/articles/alibaba-stock-and-tencent-are-tumbling-this-time-blame-the-metaverse-51645455111.

Deshmukh, Ashish (11 Sept. 2022). Intellectual Property Rights in the Metaverse. https://ijclp.com/intellectual-property-rights-in-the-metaverse/.

Des Moines Register (2014). Harvest of Change. https://www.desmoinesregister.com/pages/interactives/harvest-of-change/.

Diakopoulos, Nicholas (2019). *Automating the News: How Algorithms are Rewriting the Media*. Harvard University Press.

Diakopoulos, Nicholas (2019a). Towards a design orientation on algorithms and automation in news production. *Digital Journalism*, 7, 1180–1184.

Di Salvo, Mat (23 Feb 2023). Colombia Just Held a Court Hearing in the Metaverse—Cartoon Avatars and All. https://decrypt.co/122052/colombia-court-hearing-metaverse.

Dooley, Donovan (2023). Clark Atlanta University Gets $11.8 Million Grant To Bring Students to Metaverse. https://cassiuslife.com/569234/clark-atlanta-metaverse-grant/.

Doykos, Bernadette (Feb 21, 2022). LaMelo Ball Pulls Up to the Metaverse. https://boardroom.tv/lamelo-ball-trademark-filing-metaverse/.

Doyle, Patrick, Mitch Gelman, Sam Gill (03/16). "State of VR in Journalism." Gannett. https://www.businesswire.com/news/home/20151006006734/en/Gannett-NYTVF-Join-Forces-Present-StoryNEXT-Virtual.

Duan, H., Li, J., Fan, S., Lin, Z., Wu, X., & Cai, W. (2021). Metaverse for social good: A university campus prototype. In Proceedings of the 29th ACM International Conference on Multimedia (pp. 153–161). doi: 10.1145/3474085.3479238.

Dunne, Finley Peter (2002). Observations by Mr. Dooley. Project Gutenberg. https://www.gutenberg.org/files/4729/4729-h/4729-h.htm.

Dunlop, Douglas Morton (1943). "Muḥammad b. Mūsā al-Khwārizmī". *The Journal of the Royal Asiatic Society of Great Britain and Ireland*. 2 (3–4): 248–250.

Duval, Kathleen (2016). *Independence Lost: Lives on the Edge of the American Revolution*. New York: Random House Publishing Group.

Eady, G., Paskhalis, T., Zilinsky, J. et al. (2023). Exposure to the Russian Internet Research Agency foreign influence campaign on Twitter in the 2016 US election and its relationship to attitudes and voting behavior. Nat Commun 14, 62. https://doi.org/10.1038/s41467-022-35576-9.

Easton, Lauren (February 17, 2023). AP to develop 5 AI projects with local newsrooms. Editor and Publisher. The Associated Press. https://www.editorandpublisher.com/stories/ap-to-develop-5-ai-projects-with-local-newsrooms,242303.

Eckert, Daniel (8 July 2020. Ready, set, go: VR training hits the ground running. https://www.pwc.com.au/digitalpulse/virtual-reality-vr-training.html.

Edwards, Benj (Jan 9, 2023). Microsoft's new AI can simulate anyone's voice with 3 seconds of audio. Text-to-speech model can preserve speaker's emotional tone and acoustic environment. Ars Technica. https://apple.news/AbpoueohRSgSwV4BxM_gg1Q.

Edwards, Charlotte (4 April 2022). Police may need to patrol metaverse to stop human trafficking, expert warns. https://www.the-sun.com/tech/5049773/human-trafficking-metaverse-virtual-reality/.

Egliston, B., & Carter, M. (2022). 'The metaverse and how we'll build it': The political economy of Meta's Reality Labs. *New Media & Society*, 0(0). https://doi.org/10.1177/14614448221119785).

Eisen, Ben (9 Dec. 2022). "I Entered a Pun Competition. My Jokes Were Written by an AI Chatbot." WSJ.com https://www.wsj.com/articles/chatgpt-ai-chatbot-punderdome-jokes-11670602696.

Elgammal, A. & M. Mazzone (2019). "Art, Creativity and the Potential of Artificial Intelligence", Arts 8, 26. Special issue on The Machine as Artist (for the 21st Century).

Eliot, Lance (24 Aug. 2022). AI Ethics Lucidly Questioning This Whole Hallucinating AI Popularized Trend That Has Got To Stop. https://www.forbes.com/sites/

lanceeliot/2022/08/24/ai-ethics-lucidly-questioning-this-whole-hallucinating-ai-popularized-trend-that-has-got-to-stop/?sh=1855003777df.

Emblematic Group (2016). After Solitary. https://emblematicgroup.com/experiences/solitary-confinement/.

E-obyrnemulligan (2 Mar. 2023). How to get Spotify DJ. https://www.msn.com/en-us/news/technology/how-to-get-spotify-dj-the-new-ai-feature-explained-and-what-to-do-if-it-s-not-showing-up/ar-AA189aQq.

Eversden, Andrew (April 12, 2022). INTO THE MILITARY METAVERSE: An empty buzzword or a virtual resource for the Pentagon?—Breaking Defense. https://breakingdefense.com/2022/04/into-the-military-metaverse-an-empty-buzzword-or-a-virtual-resource-for-the-pentagon/.

Farhi, Paul (January 17, 2023). A news site used AI to write articles. It was a journalistic disaster. https://www.washingtonpost.com/media/2023/01/17/cnet-ai-articles-journalism-corrections/.

Felix & Paul Studios (2019). Traveling While Black. https://www.oculus.com/experiences/go/1994117610669719/;https://www.oculus.com/experiences/quest/2121787737926354/.

Flowers, Joe (1995). The New Scientist "How to Build a Metaverse." https://www.newscientist.com/article/mg14819994-000-how-to-build-a-metaverse/.

Flyverbom, M., Leonardi, P., Stohl, C., & Stohl, M. (2016). Digital age| the management of visibilities in the digital age—introduction. *International Journal of Communication*, 10, 12.

Force Field Entertainment B.V. (21 May 2019). National Geographic's *ExploreVR*. https://www.oculus.com/experiences/quest/2046607608728563/.

Ford, Martin (2015). *The Rise of the Robots*. New York: Basic Books.

Foti, Laurie (31 January 2022). "How Facebook Is Morphing Into Meta." https://worldnewsera.com/news/entrepreneurs/how-facebook-is-morphing-into-meta/.

Fox, Jesse, Jeremy N. Bailenson, Liz Tricase (2013). The embodiment of sexualized virtual selves: The Proteus effect and experiences of self-objectification via avatars. *Computers in Human Behavior*, Volume 29, Issue 3, Pages 930–938, ISSN 0747-5632, https://doi.org/10.1016/j.chb.2012.12.027. https://www.sciencedirect.com/science/article/pii/S074756321200369X.

Fried, Ian (24 Feb. 2022). Meta plans AI-driven universal translator. Axios: https://apple.news/ACsrhMMz4QPKm8aCZ9fh_hQ.

——(18 February 2022). Giving the VR Olympics another chance. https://www.axios.com/vr-olympics-another-chance-b5592541-ff1e-49a4-8632-90de59641804.html.

Gabora, Liane. (2013). Research on Creativity. doi: 10.1007/978-1-4614-3858-8_387. In Encyclopedia of Creativity, Invention, Innovation, and Entrepreneurship (pp.1548–1558) Edition: 1st Chapter: Research on creativity. SpringerEditors: E. G. Carayannis.

Gariffo, Micheal (6 Jan. 2023). CES 2023 was huge for AR and VR. https://www.zdnet.com/article/ces-2023-was-huge-for-ar-and-vr-heres-everything-important-that-was-announced/.

Gartner (07 February 2022). Gartner Predicts 25% of People Will Spend At Least One Hour Per Day in the Metaverse by 2026. https://www.gartner.com/en/newsroom/press-releases/2022-02-07-gartner-predicts-25-percent-of-people-will-spend-at-least-one-hour-per-day-in-the-Metaverse-by-2026.

Garton, A. (2022). How many people play Fortnite? Player count in 2022. *Dexerto*. https://www.dexerto.com/fortnite/how-many-people-play-fortnite-player-count-1666278/.

Gershgorn, Dave (15 July 2021). New Anthony Bourdain documentary deepfakes his voice: 'We can have a documentary-ethics panel about it later'. https://www.theverge.com/2021/7/15/22578707/anthony-bourdain-documentary-deepfake-voice.

Gibbs, Raymond W. (2005). *Embodiment and Cognitive Science: Concepts*. Cambridge University Press.

Gibson, James J. (1966, 1977). The theory of affordances. *Lawrence Erlbaum, Hillsdale, NJ*, 1(2).

——— (1966). The Senses Considered as Perceptual Systems.

Gilbert, Daniel (3 March 2023). The race to beat Elon Musk to put chips in people's brains. https://www.washingtonpost.com/business/2023/03/03/brain-chips-paradromics-synchron/.

Gillespie, Tarleton (2010). The Politics of 'Platforms' (May 1, 2010). *New Media & Society*, Vol. 12, No. 3, Available at SSRN: https://ssrn.com/abstract=1601487.

GitHub (n.d.a). The Immersive Web Working Group/Community Group. https://immersive-web.github.io/.

——— (n.d.b). WebXR. https://immersive-web.github.io/webxr-samples/.

Glivinska, Anna (12 Jan, 2023). The 25 Best Data Visualizations of 2023. https://visme.co/blog/best-data-visualizations/.

Go, S. Y., Jeong, H. G., Kim, J. I., & Sin, Y. T. (2021). Concept and development direction of metaverse. *Korea Information Processing Society Review*, 28(1), 7–16.

Goode, Lauren (5 May 2022). This VR App Has Legs. Spatial adds support for full-body virtual avatars, giving realism in VR a step up. WIRED: https://apple.news/AZqI715PGTB-7oELghLGQXQ.

Gosthipaty, Aritra Roy and Ritwik Raha (9 August 2021). 3D volumetric rendering with NeRF. https://keras.io/examples/vision/nerf/.

GPT-3 (21 November 2021). A Robot Wrote this Book Review. *The New York Times*. https://www.nytimes.com/2021/11/21/books/review/the-age-of-ai-henry-kissinger-eric-schmidt-daniel-huttenlocher.html?referringSource=articleShare.

Green, N and Works, K (2022). Defining the Metaverse through the lens of academic scholarship, news articles, and social media. In: Proceedings of the 27th International Conference on 3D Web Technology (Web3D '22). Article 10: 1–5. New York, NY, USA: Association for Computing Machinery.

Greenwald, Will (March 11, 2022). Elden Ring Is What the Metaverse Should Be. https://www.pcmag.com/opinions/elden-ring-is-what-the-metaverse-should-be.

Grother, Patrick, Mei Ngan and Kay Hanaoka (19 December 2019). NIST Study Evaluates Effects of Race, Age, Sex on Face Recognition Software. NIST. https://www.nist.gov/news-events/news/2019/12/nist-study-evaluates-effects-race-age-sex-face-recognition-software; https://nvlpubs.nist.gov/nistpubs/ir/2019/NIST.IR.8280.pdf.

Gugeux (20 Feb. 2022). Metaverse is not new: it is a virtual derivative to the real world. https://medium.com/@gugeux_7615/metaverse-is-not-new-it-is-a-virtual-derivative-to-the-real-world-26c57b75b22e.

Gupta, Anuprya (2 June 2015). History of Algorithmic Trading, HFT and News Based Trading. https://blog.quantinsti.com/history-algorithmic-trading-hft/.

Gures, Murat and John Yoon (Feb. 6, 2023). In Gaziantep, the epicenter of the quake, news of the devastation spread by cellphone before dawn. https://www.nytimes.com/2023/02/06/world/europe/turkey-earthquake-epicenter-gaziantep-damage.html?smid=nytcore-ios-share&referringSource=articleShare.

Haapanen, L. (2022). Problematizing the restoration of trust through transparency: Focusing on quoting. *Journalism*, 23(4), 875–891. https://doi.org/10.1177/1464884920934236.

Hacking, Mark (27 March 2022). 4 VR Headsets That Will Help You Navigate the Metaverse in Style. https://robbreport.com/gear/electronics/gallery/4-virtual-reality-headsets-1234668685/.

Haddad, Mohammed (15 Mar 2023). How does GPT-4 work and how can you start using it in ChatGPT? https://www.aljazeera.com/news/2023/3/15/how-do-ai-models-like-gpt-4-work-and-how-can-you-start-using-it.

Hamada, Rachel (6 Feb. 2023). Growing community-led investigations: the Bureau's new pilot project—The Bureau of Investigative Journalism (en-GB). https://www.thebureauinvestigates.com/blog/2023-02-06/growing-community-led-investigations-the-bureaus-new-pilot-project.

Hampson, Rick (21 Aug. 2019). How an accidental encounter brought slavery to the United States. https://www.usatoday.com/in-depth/news/nation/2019/08/21/american-slavery-began-1619-project-documents-brutal-journey/1968793001/.

Han, H. W. (2008). A study on typology of virtual world and its development in metaverse. *Journal of Digital Contents Society*, 9(2), 317–323.

Han, M., Z. Zhang, Z. Jiao, X. Xie, Y. Zhu, S. -C. Zhu & H. Liu, "Reconstructing Interactive 3D Scenes by Panoptic Mapping and CAD Model Alignments," 2021 IEEE International Conference on Robotics and Automation (ICRA), Xi'an, China, 2021, pp. 12199–12206, doi: 10.1109/ICRA48506.2021.9561546.

Handley, L. (2019). Nearly three quarters of the world will use just their smartphones to access the internet by 2025. CNBC. https://www.cnbc.com/2019/01/24/smartphones-72percent-of-people-will-use-only-mobile-for-internet-by-2025.html.

Hanson, Nick (1 Dec. 2020). Civilisations AR. https://www.bbc.co.uk/taster/pilots/civilisations-ar.

Hao, Karen (Jan. 8, 2023). China, a Pioneer in Regulating Algorithms, Turns Focus to Deepfakes. The Wall Street Journal. https://www.wsj.com/articles/china-a-pioneer-in-regulating-algorithms-turns-its-focus-to-deepfakes-11673149283.

Hardy, James (20 January 2022). The First Cell Phone: A Complete Phone History from 1920–Present. https://historycooperative.org/first-cell-phone/.

Hasan, Mohamed (18 May 2022). State of IoT 2022: Number of connected IoT devices growing 18% to 14.4 billion globally. https://iot-analytics.com/number-connected-iot-devices/).

Hassan, Jennifer, Ruby Mellen and Adam Taylor (8 March 2023). Governments around the world have moved to ban or restrict TikTok amid security fears. https://www.washingtonpost.com/world/2020/08/03/its-not-just-united-states-thesegovernments-see-tiktok-growing-problem/

Hatmaker, Taylor (February 17, 2023). Roblox wants to let people build virtual worlds just by typing. https://techcrunch.com/2023/02/17/roblox-studio-generative-ai/.

Hayden, Scott (21 March 2022a). Qualcomm Launches $100M Fund to Help Build the Metaverse; https://www.roadtovr.com/qualcomm-launches-100m-fund-help-build-metaverse/.

—— (22 March 2022b). 'Zenith: The Last City' Studio Closes $35M Series B, Aims to Create Metaverse Platform.

Hayward, Andrew (4 Feb. 2022). Nintendo Sees 'Great Potential' in the Metaverse—But It's in No Rush, Says President. Decrypt.

HBO (2022). We Met in VR. https://www.hbo.com/movies/we-met-in-virtual-reality.

Healy, Patrick, Graham Roberts, Cornelius Schmid and Yuliya Parshina-Kottas (29 Jan. 2016). "Experiencing the Presidential Campaign." http://www. nytimes.com/2016/01/30/us/politics/election-2016-virtual-reality-vr-video. html?smprod=nytcore-iphone&smid=nytcore-iphone-share&_r=0.

Heath, Alex (17 February 2022). Meta's social VR platform Horizon hits 300,000 users: That's a 10x increase in about three months. But can the growth continue? https:// www.theverge.com/2022/2/17/22939297/meta-social-vr-platform-horizon-300000-users.

—— (23 March 2022). Snap buys brain-computer interface startup for future AR glasses. https://www.theverge.com/2022/3/23/22991667/snap-buys-nextmind-brain-computer-interface-spectacles-ar-glasses.

Heckman, Chris (8 August 2021). "What is First Person Point of View?" https://www. studiobinder.com/blog/what-is-first-person-point-of-view-definition/.

Hector, Hamish (11 June 2022). The metaverse could actually be beautiful—pay attention, Meta. https://www.techradar.com/features/meta-could-learn-a-lot-from-the-beautiful-metaverse-already-on-steam.

Herman, Edward S.; McChesney, Robert W. (August 27, 2001). Global Media: The New Missionaries of Global Capitalism. A&C Black. ISBN 978-0-8264-5819-3.

Herman, Edward, and Chomsky, Noam (1988). Manufacturing Consent, New York: Pantheon Books. https://cognitive-liberty.online/chomsky-herman-propaganda-model/.

Herrman, John and Kellen Browning (July 10, 2021). Are We In The Metaverse Yet? https://www.nytimes.com/2021/07/10/style/metaverse-virtual-worlds.html.

Hernandez, Roberto, Scott Likens, Dan Priest, George Korizis, Vikram Panjwani & Emmanuelle Rivet (8 Dec. 2022). PwC 2022 US Metaverse Survey. https://www. pwc.com/us/en/tech-effect/emerging-tech/metaverse-survey.html.

Herrera, F., Bailenson, J., Weisz, E., Ogle, E., & Zaki, J. (2018). Building long-term empathy: A large-scale comparison of traditional and virtual reality perspective-taking. *PLOS ONE*, 13(10), e0204494.

Hertzfeld, Laura (5 June 2020). A Guide to AR/XR in Journalism. https://journalists. org/resources/a-guide-to-ar-xr-in-journalism/.

Holland, Patrick (14 Jan. 2022). The world in your pocket: How smartphones will get smarter in 2022. https://www.cnet.com/tech/mobile/the-world-in-your-pocket-how-smartphones-get-smarter-in-2022/.

Hollerer, T., S. Feiner and J. Pavlik, "Situated documentaries: embedding multimedia presentations in the real world," Digest of Papers. Third International Symposium on Wearable Computers, San Francisco, CA, USA, 1999, pp. 79–86, doi: 10.1109/ISWC.1999.806664.

Hsu, Tiffany and Stuart A. Thompson (Feb. 8, 2023). NYTimes: Disinformation Researchers Raise Alarms About A.I. Chatbots. https://www.nytimes. com/2023/02/08/technology/ai-chatbots-disinformation.html?smid=nytcore-ios-share&referringSource=articleShare Disinformation Researchers Raise Alarms About A.I. Chatbots.

Huang, J., & Jung, Y. (2022). Perceived Authenticity of Virtual Characters Makes the Difference. Frontiers in Virtual Reality, 178.

Huang, Kalley (Feb. 17, 2023). Microsoft to Limit Length of Bing Chatbot Conversations. https://www.nytimes.com/2023/02/17/technology/microsoft-bing-chatbot-limits. html?smid=nytcore-ios-share&referringSource=articleShare.

Huberman (Liskov), Barbara Jane (1968). A program to play chess end games, Stanford University Department of Computer Science, Technical Report CS 106, Stanford Artificial Intelligence Project Memo AI-65.

Huiyue Wu, Tong Cai, Dan Luo, Yingxin Liu, Zhian Zhang (2021). Immersive virtual reality news: A study of user experience and media effects. *International Journal of Human-Computer Studies*, Volume 147, 102576, ISSN 1071-5819, https://doi.org/10.1016/j.ijhcs.2020.102576. https://www.sciencedirect.com/science/article/pii/S1071581920301786.

Hunter, Tatum (18 Jan. 2023). ChatGPT could make life easier. Here's when it's worth it. https://www.washingtonpost.com/technology/2023/01/18/chatgpt-personal-use/.

—— (April 2022). Let's get Meta-physical: Why Oculus fitness actually works. https://www.washingtonpost.com/technology/2022/04/21/vr-workout-games/.

IANS (18 February 2022). Over 1 in 3 consumers never heard of Metaverse: Report. https://www.timesnownews.com/technology-science/article/over-1-in-3-consumers-never-heard-of-Metaverse-report/860051.

IBM Cloud Education (3 June 2020). Artificial Intelligence. https://www.ibm.com/cloud/learn/what-is-artificial-intelligence.

IBM Research (6 November 2013). Watson and the Jeopardy! Challenge. https://www.youtube.com/watch?v=P18EdAKuC1U.

IMDB (2022). We Met in Virtual Reality. https://m.imdb.com/title/tt16378482/.

Internet Archive (2023). https://archive.org/details/you-are-there-1950-04-16-87-thermopolae.

Internet Society (2017). "History of Virtual Reality." https://www.vrs.org.uk/virtual-reality/history.html.

—— (2017). "VPL Research Jaron Lanier." https://www.vrs.org.uk/virtual-reality-profiles/vpl-research.html.

Irwin, Kate (25 Feb. 2022). Microsoft's Minecraft Goes Web3 With 'NFT Worlds' on Polygon. https://decrypt.co/93783/microsoft-minecraft-web3-nft-worlds-ethereum-polygon.

Iyengar, Shanto and Donald R Kinder (2010). *News that matters: Television and American opinion*. University of Chicago Press.

Jamieson, K. H. (2012). Does the US Media Have a Liberal Bias. *Perspective on Politics*, 10 (3), 783–785. https://doi.org/10.1017/S1537592712001193.

Jang, J. (2021). A study on a Korean speaking class based on metaverse: Using Gather.town. *J. Korean Lang. Educ*, 32, 279–301.

Jarvey, Natalie (18 April 2022). How Snapchat's AR lenses are helping Netflix, Sony, and more studios tap into the metaverse to market films from 'Morbius' to 'Marry Me'. https://www.businessinsider.com/snapchat-ar-lens-netflix-sony-metaverse-movie-marketing-morbius-2022-4.

Jennings, Ralph (12 April 2022). Chinese Computer Giant Lenovo Plans A Big Push Into The Metaverse. https://www.forbes.com/sites/ralphjennings/2022/04/12/chinese-computer-giant-lenovo-plans-a-big-push-into-the-metaverse/.

Jeon, J. H. (2021). A study on education utilizing metaverse for effective communication in a convergence subject. *International Journal of Internet Broadcasting and Communication*, 13(4), 129–134.

Johnson, Steven (15 April 2022). A.I. Is Mastering Language. Should We Trust What It Says? https://www.nytimes.com/2022/04/15/magazine/ai-language.html?referringSource=articleShare.

Joire, Myriam (March 27, 2022). I played VR games in a moving car and didn't get sick: We test Holoride's immersive in-car VR experiences (uses vive). https://global. techradar.com/en-za/news/i-played-vr-games-in-a-moving-car-and-didnt-get-sick.

Jones, Steven E. (2006). *Against Technology: From the Luddites to Neo-Luddism*. Routledge. ISBN 978-0-415-97868-2.

Jovanović, A., & Milosavljević, A. (2022). VoRtex Metaverse platform for gamified collaborative learn- ing. *Electronics (Basel)*, 11(3), 317. doi: 10.3390/electronics11030317.

JPMorgan Chase (13 February 2022). TheStreet Quant Ratings. https://www.thestreet. com/files/r/ratings/equities/JPM_weiss.pdf.

Kahn, Roomy (18 February 2022). Metaverse: Enhancing Life Experiences Beyond The Physical And Temporal Boundaries. https://www.forbes.com/sites/ roomykhan/2022/02/18/Metaverse-enhancing-life-experiences-beyond-the-physical-and-temporal-boundaries/?sh=7d3ab98f1af8.

Kaimara, P., Oikonomou, A., & Deliyannis, I. (2022). Could virtual reality applications pose real risks to children and adolescents? *Virtual Reality*, 26(2), 697–735.

Kamin, Debra (19 Feb. 2023). The Next Hot Housing Market Is Out of This World. It's in the Metaverse. https://www.nytimes.com/2023/02/19/realestate/metaverse-vr-housing-market.html?smid=nytcore-ios-share&referringSource=articleShare.

Kanematsu, H., Kobayashi, T., Barry, D. M., Fukumura, Y., Dharmawansa, A., & Ogawa, N. (2014). Virtual STEM class for nuclear safety education in metaverse. *Procedia Computer Science*, 35, 1255–1261. doi: 10.1016/j.procs.2014.08.224.

Kawamoto, D. (17 February 2011). Watson Wasn't Perfect: IBM Explains the 'Jeopardy!' Errors. AOL.com https://www.aol.com/2011/02/17/the-watson-supercomputer-isnt-always-perfect-you-say-tomato/.

Kealey, Kate (17 July 2022). "Iowa Company Creates Virtual Reality Classrooms for 10 Universities." https://www.the74million.org/article/iowa-company-creates-virtual-reality-classrooms-for-10-universities/.

Keefe, John (19 Sept. 2017). Quartz's iPhone app now includes news stories in augmented reality. https://qz.com/1072252/quartzs-iphone-app-now-includes-news-stories-in-augmented-reality.

Kelly, Jack (31 January 2022). "Sandbox VR Offers An Immersive, Interactive, Blockbuster-Action Movie, Virtual-Reality Experience." https://www.forbes.com/sites/jackkelly/2022/01/31/sandbox-vr-offers-an-immersive-interactive-blockbuster-action-movie-virtual-reality-experience/.

Kennicott, Philip (May 15, 2017). A new concert hall in Hamburg transforms the city. http://wapo.st/2FxwYRS.

Khaleej Times Staff (17 June 2022). Metaverse is a science-fact, says Microsoft VP. Khaleej Times. https://www.khaleejtimes.com/tech/metaverse-is-a-science-fact-now-says-microsoft-vp.

Kickstarter Design & Tech (April 4, 2022). Reach Out and Touch the Metaverse with Your Bare Hands—Core77. https://www.core77.com/posts/114737/Reach-Out-and-Touch-the-Metaverse-with-Your-Bare-Hands.

Kieslich, Kimon, Pero Došenović, Christopher Starke, Marco Lünich, & Frank Marcinkowski (January 2021). Artificial Intelligence in Journalism. Meinungsmonitor Künstliche Intelligenz.

Kilteni, K., Groten, R., & Slater, M. (2012). The sense of embodiment in virtual reality. *Presence: Teleoperators and Virtual Environments*, 21(4):373–387, 2012.

Kim, J. G. (2021). A study on metaverse culture contents matching platform. *International Journal of Advanced Culture Technology*, 9(3), 232–237.

Kim, Taeyong and Frank Biocca (23 June 2006). "Telepresence via Television." *Journal of Computer-Mediated Communication*. https://onlinelibrary.wiley.com/doi/full/10.1111/j.1083-6101.1997.tb00073.x.

Kimball, Whitney (9 June 2021). The Future Is NFTing Doodles in Minutes. Gizmodo. https://gizmodo.com/the-future-is-nfting-doodles-in-minutes-1847061360.

King, Hope (1 January 2024). Davos' metaverse access. Axios. https://www.axios.com/2024/01/15/davos-metaverse-global-collaboration-village.

King, Steven (12 Mar. 2018). Augmenting the World with News. https://steven-king.medium.com/sxsw-panel-reality-but-better-augmenting-the-world-with-news-899d3cc3b8af.

Kirkpatrick, N., Atthar Mirza and Manuel Canales (27 March 2023). The blast effect. https://www.washingtonpost.com/nation/interactive/2023/ar-15-damage-to-human-body/.

Kissinger, Henry A, Eric Schmidt, and Daniel Huttenlocher (2021). The Age of AI and Our Human Future. Little, Brown.

Knight, Shawn (2 March. 2022). HTC to launch first 'metaverse' phone in April. TechSpot. https://www.techspot.com/news/93624-htc-launch-first-metaverse-phone-april.html.

Knight, Will (14 March 2023). GPT-4 Will Make ChatGPT Smarter but Won't Fix Its Flaws. https://www.wired.com/story/gpt-4-openai-will-make-chatgpt-smarter-but-wont-fix-its-flaws/.

Koch, Andreas, Ingolf Cascorbi, Martin Westhofen, Manuel Dafotakis, Sebastian Klapa, & Johann Peter Kuhtz-Buschbeck (12 October 2018). The Neurophysiology and treatment of motion sickness. *Deutsches Ärzteblatt International*, 115(41):687–696. doi: 10.3238/arztebl.2018.0687. PMID: 30406755; PMCID: PMC6241144.

Koetsler, John (8 November 2022). Generative AI: The Future Is AI Writing Its Own Code. Forbes.com https://www.forbes.com/sites/johnkoetsier/2022/11/08/generative-ai-the-future-is-ai-writing-its-own-code/?sh=570e96de1bd0.

Koliska, Michael. (2015). Transparency and trust in journalism. University of Maryland, College Park ProQuest Dissertations Publishing, 2015. 3726215. https://doi.org/10.1093/acrefore/9780190228613.013.883.

Kolodny, Lora (7 Nov. 2022). Elon Musk bans impersonation without parody label on Twitter raising questions about free speech commitment. https://www.cnbc.com/2022/11/07/elon-musk-unlabeled-twitter-parody-accounts-risk-permanent-suspension.html.

Konnanath, Gopikrishnan (31 March 2022). How to build the metaverse with AR, VR, and blockchain. https://www.businesstoday.in/opinion/columns/story/how-to-build-the-metaverse-with-ar-vr-and-blockchain-328056-2022-03-31.

Koppel, Niko, Nick Capezzera & Veda Shastri (31 May 2017). Life on Mars. https://www.nytimes.com/interactive/2017/05/31/science/space/life-on-mars.html.

Kovach, Bill and Rosenstiel, Tom (2007). *The Elements of Journalism: What Newspeople Should Know and the Public Should Expect, Completely Updated and Revised*. Three Rivers Press.

Kye, B., Han, N., Kim, E., Park, Y., & Jo, S. (2021). Educational applications of metaverse: Possibilities and limitations. *Journal of Educational Evaluation for Health Professions*, 18. doi: 10.3352/jeehp.2021.18.32 PMID:34897242.

Kyrlitsias, Christos, & Despina Michael-Grigoriou (2018). Asch conformity experiment using immersive virtual reality. *Computer Animation & Virtual Worlds*. First published: 12 March 2018. https://doi.org/10.1002/cav.1804.

Lai, Richard (22 January 2024). "Apple might have sold up to 180,000 Vision Pro headsets over pre-order weekend." Engadget. https://www.engadget.com/apple-might-have-sold-up-to-180000-vision-pro-headsets-over-pre-order-weekend-081727344.html?guccounter=1.

Lang, Ben (13 April 2022). Pokémon Go' Studio's New Game, 'Peridot', Leans Further into AR Gameplay. https://www.roadtovr.com/niantic-peridot-announcement-ar-gameplay-focus/.

—— (5 Oct. 2022). Meta's First Step Toward a Proper Metaverse: Avatars That Can "travel between" Two Platforms. https://www.roadtovr.com/meta-first-steps-toward-metaverse-avatar-travel-horizon-crayta/.

Lanier, Jaron (2013). *Who Owns the Future*. Simon and Schuster.

—— (3 February 2022). *Virtual Seminar on Virtual Reality*. Columbia Institute for Tele-Information.

Lasswell, Harold (1933). *'Propaganda,' Encyclopaedia of the Social Sciences*, New York: Macmillan, pp. 521–28.

LATimes (24 May 2019). What is the Quakebot and how does it work? https://www.latimes.com/la-me-quakebot-faq-20190517-story.html.

Lavenda, David (22 February 2022). The New Metaverse: What's Different This Time? https://www.cmswire.com/digital-workplace/the-new-Metaverse-whats-different-this-time/.

Laws, Ana Luisa Sánchezn (2019). *Conceptualising Immersive Journalism*. Routledge.

Lba, Alejandro (10/22/15). "New York Times launches VR app, delivers 1 million Google Cardboards to print subscribers." http://www.nydailynews.com/news/national/ny-times-launches-vr-app-ships-1-million-google-cardboards-article-1.2407586.

Lee, H., & Hwang, Y. (2022). Technology-Enhanced Education through VR-Making and Metaverse-Linking to Foster Teacher Readiness and Sustainable Learning. *Sustainability*, 14(8), 4786. doi: 10.3390u14084786.

Lee, S. (2020). *Log in Metaverse: Revolution of Human Space Time*. Issue Report.

Lemmens, Jeroen S., Monika Simon, & Sindy R. Sumter. (2022). Fear and loathing in VR: the emotional and physiological effects of immersive game. *Virtual Reality*, 26(1): 223–234. doi: 10.1007/s10055-021-00555-w.

Lenihan, Rob (19 February 2022). Why Companies Are Flocking to The Metaverse. https://www.miamiherald.com/opinion/op-ed/article255429986.html.

Levins, Hoag (2000). APB Wins Two NYC Press Club Awards. http://www.levins.com/apbaward05.html.

Levy, Mark R. (December 1982). The Lazarsfeld-Stanton Program Analyzer: An Historical Note. *Journal of Communication*, Volume 32, Issue 4, Pages 30–38, https://doi.org/10.1111/j.1460-2466.1982.tb02516.x.

Likens, Scott & Andrea Mower. (15 Sept. 2022). What does virtual reality and the metaverse mean for training? https://www.pwc.com/us/en/tech-effect/emerging-tech/virtual-reality-study.html.

Lima, Cristiano & David DiMolfetta (17 March 2023). AI chatbots won't enjoy tech's legal shield, Section 230 authors say. https://www.washingtonpost.com/politics/2023/03/17/ai-chatbots-wont-enjoy-techs-legal-shield-section-230-authors-say/.

Lindner, P., Miloff, A., Zetterlund, E., Reuterskiöld, L., Andersson, G., & Carlbring, P. (2019). Attitudes toward and familiarity with virtual reality therapy among practicing cognitive behavior therapists: a cross-sectional survey study in the era of consumer VR platforms. *Frontiers in psychology*, 10, 176.

Locatelli, Luca (07/22/16). "Pilgrimage: A 21st-Century Journey to Mecca and Medina." https://www.nytimes.com/2016/07/22/world/middleeast/pilgrimage-virtual-reality-in-mecca-and-medina.html.

Lochtefeld, James (2002). "Avatar" in The Illustrated Encyclopedia of Hinduism, Vol. 1: A-M, Rosen Publishing, ISBN 0-8239-2287-1, pages 72–73.

Lombard, Matthew, Teresa B. Ditton and Lisa Weinstein (2010). "Measuring Presence: The Temple Presence Inventory." https://www.researchgate.net/publication/228450541_Measuring_Presence_The_Temple_Presence_Inventory 7, p. 222.

Lu, Kristine and Katerina Eva Matsa (8 September 2016). More than half of smartphone users get news alerts, but few get them often. https://www.pewresearch.org/fact-tank/2016/09/08/more-than-half-of-smartphone-users-get-news-alerts-but-few-get-them-often/.

Lucia, B., Vetter, M.A. and Adubofour, I.K. (2023). Behold the metaverse: Facebook's Meta imaginary and the circulation of elite discourse. *New Media & Society*. Epub ahead of print 10 July 2023. https://doi.org/10.1177/14614448231184249.

Lutz, Sander (30 March 2023). Are DAOs Dead? https://decrypt.co/124963/daos-judge-ruling-bzx-case-trouble.

Lyons, Kim (31 January 2022). "Meta adds 3D avatars to Instagram Stories, with updates for Messenger and Facebook." https://www.theverge.com/2022/1/31/22910271/meta-3d-avatars-instagram-stories-messenger-Facebook.

Maneli, M.A. and O. E. Isafiade. (2022). "3D Forensic Crime Scene Reconstruction Involving Immersive Technology: A Systematic Literature Review," in *IEEE Access*, vol. 10, pp. 88821–88857, doi: 10.1109/ACCESS.2022.3199437.

Maese, Rick, Madison Walls, Artur Galocha, Leslie Shapiro and Ashleigh Joplin (20 July 2021). Sporting Climbing. WashingtonPost.com. https://www.washingtonpost.com/sports/olympics/interactive/2021/sport-climbing-brooke-raboutou-augmented-reality/.

Mahadevan, Alex (February 3, 2023). This newspaper doesn't exist: How ChatGPT can launch fake news sites in minutes—Poynter. https://www.poynter.org/fact-checking/2023/chatgpt-build-fake-news-organization-website/.

Majumder, Bhaswati Guha (26 March 2022). From TechMVerse to Metaverse in India. https://www.news18.com/news/tech/exclusive-from-techmverse-to-metaverse-in-india-tech-mahindra-official-explains-it-all-4911260.html.

Maritain, J. (1953). Creative Intuition in Art and Poetry. Bollingen Ser 35; New York.

Marr, Bernard (23 August 2019). The Amazing Ways YouTube Uses Artificial Intelligence And Machine Learning. https://www.forbes.com/sites/bernardmarr/2019/08/23/the-amazing-ways-youtube-uses-artificial-intelligence-and-machine-learning/?sh=59db20a85852.

Marson, James (29 November 2018). The Future of journalism: News games. JLDN. https://journalism.london/2018/11/the-future-of-journalism-news-games/.

Martingano, A. J., Hererra, F., & Konrath, S. (2021). Virtual Reality Improves Emotional but Not Cognitive Empathy: A Meta-Analysis. *Technology, Mind, and Behavior*, 2(1). https://doi.org/10.1037/tmb0000034.

McCoy, Terrence (February 20, 2023). The Amazon, Undone: Death in the Forest. WashingtonPost.com. https://www.washingtonpost.com/world/interactive/2022/brazil-amazon-deforestation-highway-br-319/?itid=lk_inline_manual_14.

McKeown, Kathleen, Regina Barzilay, John Chen, David Elson, David Evans, Judith Klavans, Ani Nenkova, Barry Schiffman and Sergey Sigelman (2003). Columbia's Newsblaster. In Proceedings of NAACL-HLT'03 https://www.cs.columbia.edu/nlp/papers/2003/mckeown_al_03a.pdf.

Mellon, Rory (11 March 2023). I took a $599 gamble on the PSVR 2—and it paid off. https://www.tomsguide.com/features/i-took-a-dollar550-gamble-on-the-psvr-2-heres-what-happened.

Melnick, Kyle (17 February 2022). VR Social Platform VRChat Adds Face Tracking For Avatars. https://vrscout.com/news/vr-social-platform-vrchat-adds-face-tracking-for-avatars/.

—— (22 March 2022). Holographic Tabletop Games On The Way From Catan Creators. https://vrscout.com/news/holographic-tabletop-games-on-the-way-from-catan-creators/.

—— (4 April 2022). Snoop Dogg's New Music Video Was Shot In The Metaverse—VRScout. https://vrscout.com/news/snoop-doggs-new-music-video-was-shot-in-the-metaverse/.

—— (April 6, 2022). Experience What It's Like To Be Slapped By Will Smith In VR. https://vrscout.com/news/experience-what-its-like-to-be-slapped-by-will-smith-in-vr/.

Mergerson, Christoph (24–28 May 2018). "Legal implications of sexual assault and rape in virtual reality." 68th ICA Annual Conference, Prague, Czech Republic.

Merton, Robert K. (1936). "The unanticipated consequences of purposive social action." *American sociological review*, 1(6), 894–904.

Meta Platforms, Inc (2021) "Horizon Worlds." https://www.oculus.com/horizon-worlds/.

MetaAI (2023). Introducing LLaMA. https://ai.facebook.com/blog/large-language-model-llama-meta-ai/.

MetaQuest (2022). "The Arcade." https://www.oculus.com/vr/6608791699161643/.

Metaverse News (2022). https://metaverseinsider.tech/.

Metz, Cade and Karen Weise (7 February 2023). A Tech Race Begins as Microsoft Adds A.I. to Its Search Engine. The New York Times. https://www.nytimes.com/2023/02/07/technology/microsoft-ai-chatgpt-bing.html?smid=nytcore-ios-share&referringSource=articleShare.

Microsoft (2022). Social capital. https://www.microsoft.com/en-us/worklab/bringing-us-together.

—— (27 May 2022). What is a voice assistant? https://learn.microsoft.com/en-us/azure/cognitive-services/speech-service/voice-assistants.

Milano, Matt (April 15, 2022). Shocker: US teens don't care about the metaverse. https://www.androidauthority.com/teens-metaverse-3153930/.

Miller M.R., Bailenson J.N. (2021). "Social Presence Outside the Augmented Reality Field of View," *Frontiers in Virtual Reality*, doi: 10.3389/frvir.2021.656473.

Miller, Ross (20 October 2015). The company behind the AP's 'robot journalists' is opening up its technology for everyone. https://www.theverge.com/2015/10/20/9572975/automated-insights-wordsmith-natural-language.

Mims, Christopher (9 April 2022). The Future of Socializing at Work? Virtual Golf.

Mingay, Adam (April 19, 2022). The gateway to the metaverse isn't gaming, it's AR. https://www.thedrum.com/opinion/2022/04/19/the-gateway-the-metaverse-isnt-gaming-its-ar.

Mlot, Stephanie (January 5, 2023). Apple says digital narration for Apple Books will make it cheaper, and therefore more accessible to produce audiobooks. Voice actors and audiobook publishers may disagree. https://www.pcmag.com/news/apple-rolls-out-ai-narrated-audiobooks.

Monzon, Luis (28 Feb. 2022). MTN is the First African Company to Officially Enter the Metaverse; https://www.itnewsafrica.com/2022/02/mtn-is-the-first-african-company-to-officially-enter-the-metaverse/.

Moore, Gordon E. (19 April 1965). "Cramming more components onto integrated circuits" (PDF). *intel.com*. Electronics Magazine.

Morales, Christina (24 May 2021). 'Charlie Bit My Finger' Is Leaving YouTube After $760,999 NFT Sale. https://www.nytimes.com/2021/05/24/arts/charlie-bit-my-finger-nft-auction.html.

Mori, Masahiro (2012). Translated by Karl F. MacDorman; Norri Kageki. "The uncanny valley". IEEE Robotics and automation. *New York City: Institute of Electrical and Electronics Engineers*, 19(2): 98–100. doi: 10.1109/MRA.2012.2192811.

Moye, Jay (7 Nov. 2022). Inside the writers' room cranking out Alexa's jokes, from the cheeky to the downright corny. https://www.aboutamazon.com/news/devices/inside-the-writers-room-cranking-out-alexas-jokes-from-the-cheeky-to-the-downright-corny.

Moyer, Suzette (October 23, 2017). You've never carved a pumpkin like this before, http://wapo.st/2HlhbWI.

Mozée, Carla (30 November 2021). "A plot of virtual land that went for $4.3 million in The Sandbox is the most expensive Metaverse property sale ever." https://markets.businessinsider.com/news/currencies/Metaverse-property-sandbox-virtual-real-estate-deal-record-4-million-2021-11?op=1.

Mulqueen, Tina (MAR 23, 2022). What My Kids' Roblox Addiction Taught Me About the Metaverse. https://www.entrepreneur.com/article/420727.

Murray, Paul (15 March 2023). Who Is Still Inside the Metaverse? https://nymag.com/intelligencer/article/mark-zuckerberg-metaverse-meta-horizon-worlds.html.

Murphy, D., & Skarbez, R. (2022). What Do We Mean When We Say "Presence"?. PRESENCE: Virtual and Augmented Reality, 1–43.

Myers, Steven Lee, Paul Mozur & Jeff Kao (19 February 2022). Bots, fake accounts push China's Olympics vision https://asiatimes.com/2022/02/bots-fake-accounts-push-chinas-olympics-vision/.

Nakatani, Mina (3 Feb. 2023). The True Story of The Circulation War Between Hearst and Pulitzer. https://www.grunge.com/387285/the-true-story-of-the-circulation-war-between-hearst-and-pulitzer/.

National Press Foundation (18 Nov 2015). Des Moines Register and Gannett Product Win First Ever, 'Best Use of Technology in Journalism' Award for Virtual Reality and 360-Degree Video | MarketScreener. https://m.marketscreener.com/quote/stock/GANNETT-CO-INC-22890963/news/Gannett-Des-Moines-Register-and-Gannett-Product-Win-First-Ever-lsquo-Best-Use-of-Technology-in-J-21431162/.

Native News Online (2022). What Metaverse Projects to Watch in 2022. https://nativenewsonline.net/advertise/branded-voices/what-metaverse-projects-to-watch-in-2022.

NBC News MACH (10 July 2017). Bill Nye in VR. https://www.nbcnews.com/mach/science/bill-nye-explain-his-eight-principles-everything-vr-event-ncna781296.

NBD (11 October 2021). http://www.nbd.com.cn/corp/nbd_live/dist/index.html#/?id=1349.

Neacsu, Dana (2022). Conversation with the author.

Needleman, Sarah (19 April 2022). Friend or Bot? Phony Gamers Leave Players Feeling Betrayed. Bots disguised as people. The Wall Street Journal. https://www.wsj.com/articles/friend-or-bot-phony-gamers-leave-players-feeling-betrayed-11650377760.

Nelson, Jason (Feb. 17, 2023). What Are Ordinals? A Beginner's Guide to Bitcoin NFTs—Decrypt. https://decrypt.co/resources/what-are-ordinals-a-beginners-guide-to-bitcoin-nfts.

Nelson, Theodor Holm (August 1965). "Complex information processing." ACM '65: Proceedings of the 1965 20th National Conference. ACM: 84–100. doi: 10.1145/800197.806036. ISBN 9781450374958. S2CID 2556127.

New York Times, The (2023). Immersive https://www.nytimes.com/spotlight/augmented-reality).

—— (2023). The Daily 360. https://www.nytimes.com/video/the-daily-360.

—— (21 July 2018). Step Inside the Thai Cave in Augmented Reality. https://www.nytimes.com/interactive/2018/07/21/world/asia/thai-cave-rescue-ar-ul.html.

—— (Nov. 2022). You Be the Ump (via interactive AR). https://www.nytimes.com/interactive/2022/sports/baseball/umpire-pitch-ball-strike-game.html?smid=nytcore-ios-share&referringSource=articleShare.

News18 (17 February 2022). Qualcomm Brings Wi-Fi 7 Standard To Push Metaverse And AR Capabilities. https://www.news18.com/news/tech/qualcomm-brings-wi-fi-7-standard-to-push-Metaverse-and-ar-capabilities-4781720.html.

Nicholls, Beth (March 20, 2022). AI and AR are the future of visual storytelling, according to Canon experts; https://www.digitalcameraworld.com/news/ai-and-ar-are-the-future-of-visual-storytelling-according-to-canon-experts.

Nissenbaum, H. (2010). Privacy in Context. Stanford University Press.

Noble, Holcomb B. (December 26, 2006). "Frank Stanton, Broadcasting Pioneer, Dies at 98". The New York Times.

Norman, Jeremy (2023). History of Information. https://www.historyofinformation.com/detail.php?entryid=1087).

Norton, W. J., Sauer, J., & Gerhard, D. (2022). A quantifiable framework for describing immersion. Presence, 29, 191–200.

Notre-Dame de Paris: Augmented Exhibition (2022). https://notredameexpo.com/en/.

NowHere Media (19 Aug. 2020). Home After War. https://www.oculus.com/experiences/quest/2900834523285203/?ranking_trace=0_2900834523285203_QUESTSEARCH_b0131e0f-88cd-4a3c-b8c3-fa606b1a9803;https://www.homeafterwar.net/.

Nunez, Adriana (12 April 2022). Mastercard's metaverse ambitions take shape. https://www.emarketer.com/content/mastercard-s-nft-metaverse-related-trademark-filings-highlight-virtual-ambitions.

Ochanji, Sam (13 March 2022). OpenXR Wants to Standardize Advanced Haptics for VR and AR. http://virtualrealitytimes.com/2022/03/13/openxr-wants-to-standardize-advanced-haptics-for-vr-and-ar/.

—— (27 March 2022). Nike's Metaverse Store Has Had 7 Millions Visits So Far (NFTs and more via Roblox metaverse). http://virtualrealitytimes.com/2022/03/27/nikes-metaverse-store-has-had-7-millions-visits-so-far/.

Ohm, Paul. (March 18, 2009). The Probability of Privacy. CRCS Privacy and Security Lunch Seminar. Harvard. https://crcs.seas.harvard.edu/event/paul-ohm-probability-privacy.

Okaloosa News-Journal, The (1 Nov. 1956). "Cinerama Crews Shooting New Movie At Eglin AFB", Crestview, Florida. The Okaloosa News-Journal—Edgewater Area News section, Volume 42, Number 44, page 1.

Online News Association (10 July 2020). Immersive Case Study: Euronews 360 Video. https://journalists.org/resources/immersive-case-study-euronews-360-video/; https://youtube.com/playlist?list=PLSyY1udCyYqDLlN78Jwcark22yjdoiOpF.

OpenAI (2021). Aligning AI systems with human intent. https://openai.com.

—— (2022). "GPT-3." Wikipedia. https://en.wikipedia.org/wiki/GPT-3.

—— (2023). ChatGPT Plus. https://openai.com/blog/chatgpt-plus.

Oremus, Will (2022). "Kids are flocking to Facebook's 'Metaverse.' Experts worry predators will follow." https://www.washingtonpost.com/technology/2022/02/07/facebook-metaverse-horizon-worlds-kids-safety/.

Oruganti, Rama (02 May 2022). The Extended Reality Spectrum Delivers a Variety of Metaverse Experiences. https://readwrite.com/xr-spectrum-delivers-variety-of-metaverse-experiences/.

Ovide, Shira (March 22, 2022). Recreating the human face could give virtual interactions the sense of intimacy they're missing. https://www.nytimes.com/2022/03/22/technology/metaverse-faces-steve-perlman.html?referringSource=articleShare.

Pacheco, Dan (forthcoming, expected 2023). "Immersive Journalism" chapter in John V. Pavlik (editor) Milestones in Digital Journalism. Forthcoming from Routledge Press.

Paleja, Ameya (25 March 2022). A Japanese company wants you to feel real pain in the metaverse. Interesting Engineering. https://interestingengineering.com/innovation/japanese-company-pain-metaverse.

Pandey, Rajesh (Nov. 13, 2022). Apple is building a rival to the metaverse. https://www.cultofmac.com/796947/apple-building-rival-metaverse/.

Paananen, Villem, Mohammad Sina Kiarostami, Lik-Hang Lee, Tristan Braud, and Simo Hosio (2022). From Digital Media to Empathic Reality: A Systematic Review of Empathy Research in Extended Reality Environments. In ACM CSUR. ACM, New York, NY, USA, 35 pages. https://doi.org/10.1145/1122445.1122456.

Pantony, Ali (8 March 2022). Where Do Women Fit Into The Metaverse? https://www.glamour.com/story/where-do-women-fit-into-the-metaverse.

Parashar, Radhika and Richa Sharma (20 February 2023). YouTube Set for Web3, Metaverse Revamp Under New CEO Neal Mohan. https://www.gadgets360.com/cryptocurrency/news/youtube-web3-metaverse-revamps-under-new-ceo-neal-mohan-3798388.

Park, Danny (11 April 2023). France to develop local metaverses against global giants https://forkast.news/headlines/france-local-metaverses-against-giants/.

Pasquarelli, Walter (6 August 2019). Towards Synthetic Reality: When DeepFakes meet AR/VR. OxfordInsights. https://oxfordinsights.com/insights/towards-synthetic-reality-when-deepfakes-meet-ar-vr/).

Pastel, Stefan, Dan Bürger, Chien-Hsi Chen, Katharina Petri, & Kerstin Witte (2022). Comparison of spatial orientation skill between real and virtual environment. Virtual Reality, 26: 91–104. https://doi.org/10.1007/s10055-021-00539-w.

Patra, Ishan (27 March 2022). Meta's Oculus Quest 2 leads global AR, VR market: IDC. https://www.thehindu.com/sci-tech/technology/metas-oculus-quest-2-leads-global-ar-vr-market-idc/article65268178.ece.

Patterson, Thomas E. (2019). *How America Lost Its Mind: The Assault on Reason That's Crippling Our Democracy.* University of Oklahoma Press. ISBN 9780806164328.

Pavlik, John (2021). Experiencing Cinematic VR: Where Theory and Practice Converge in the Tribeca Film Festival Cinema360." New York State Communication Association Annual Conference, 14 October 2020, Callicoon, New York. Proceedings of the New York State Communication Association: Vol. 2020, Article 9. Available at: https://docs.rwu.edu/nyscaproceedings/vol2020/iss1/9.

—— (14–16 Oct. 2022). Entering the Metaverse. Presented at NYSCA 2022 14–16 October Callicoon, NY.

—— (2023). Detecting Bias in DALL-E Imagery. Paper for the New York State Communication Association Annual Conference, Callicoon, NY.

—— (2019). *Journalism in the Age of Virtual Reality: How Experiential Media and the Transforming News.* (New York: Columbia University Press).

Pavlik, John V. and S. Feiner (2021). Haptic journalism: a research proposal to the Tow Center for Journalism.

Pavlik, John V. and Shravan Regret Iyer. "Of Media and Mediums: Illusion and the Roots of Virtual Reality in Victorian Era Science, Social Change & Spiritualism. Presented at the New York State Communication Association (NYSCA) conference, Callicoon, New York, 15–17 October 2021.

Pearson, Jordan (23 Feb. 2022). Web3 Developers Have Found a Functioning Metaverse. It's 'Minecraft'. https://www.vice.com/en/article/wxd4w5/web3-developers-have-discovered-a-functioning-metaverse-its-minecraft.

Pearson, Luke and Sandra Youkhana (1 April 2022). The metaverse doesn't look as disruptive as it should, it looks ordinary—here's why. https://theconversation.com/the-metaverse-doesnt-look-as-disruptive-as-it-should-it-looks-ordinary-heres-why-175866.

Pelham, S. (2020, April 15). Ohio students use 360-degree videos to document daily life during COVID-19. Ohio University News. https://www.ohio.edu/news/2020/04/ohio-students-use-360-degree-videos-document-daily-life-during-covid-19.

Perplexity.ai (2023). https://www.perplexity.ai/?uuid=3025e555-1d43-4b83-a008-492019558c8c.

Pew (January 12, 2021). More than eight-in-ten Americans get news from digital devices. https://www.pewresearch.org/fact-tank/2021/01/12/more-than-eight-in-ten-americans-get-news-from-digital-devices/.

Pew Research Center (13 July 2021). Cable News Fact Sheet. https://www.pewresearch.org/journalism/fact-sheet/cable-news/.

—— (2020). Analysis of MEDIA Access Pro & BIA Advisory Services data, 2020.

Pimentel, K., & Teixeira, K. (1993). *Virtual reality.* New York, NY: McGraw-Hill. ISBN 978-0-8306-4065-2.

PKouppas (2013). News Alive—Sunday Telegraph—News Australia—Newspaper Augmented Reality. https://youtu.be/l9_rL7AP3ik.

PlayStation reveals PS VR2, the next generation of their virtual reality headset. https://www.washingtonpost.com/video-games/2022/02/22/ps-vr2-headset-sony/.

Poell, et al. (2021). Critical Augmented and Virtual Reality Network (CAVRN). From the political economy perspective, XR situated within them "platformization of cultural production" https://cavrn.org/.

Pogue, David (11 April 2022). "Crypto for dummies": The how, what and why of using virtual currency. CBS. https://www.cbsnews.com/news/cryptocurrency-bitcoin-virtual-currency-explainer/.

——— (15 January 2023). Artificial Intelligence Art created by AI. Dall-E and more. CBS Sunday Morning. https://www.cbsnews.com/news/ai-art-created-by-artificial-intelligence/.

——— (22 January 2023). AI experts on whether you should be "terrified" of ChatGPT. CBS Sunday Morning. https://www.cbsnews.com/news/ai-experts-on-chatgpt-artificial-intelligence-writing-program/.

Priestley, Jenny (March 30, 2022). Streaming v the Metaverse: Deloitte publishes 2022 Digital Media Trends report—TVBEurope. (3D meta verse over streaming). https://www.tvbeurope.com/media-consumption/streaming-v-metaverse-deloitte-publishes-2022-digital-media-trends-report.

Primrose, Jason (9 April 2022). Web3 and blockchain technology: How digital asset ownership is flipping the current business model on its head. https://venturebeat.com/2022/04/09/web3-and-blockchain-technology-how-digital-asset-ownership-is-flipping-the-current-business-model-on-its-head/.

ProjectTopics (3 May 2021). Smartphone UTILIZATION IN NEWS GATHERING AMONG JOURNALISTS. https://projectchampionz.com.ng/2021/05/03/smart-phone-utilization-in-news-gathering-among-journalist/.

Pulitzer Prizes, The (2018). The 2018 Pulitzer Prize Winner in Explanatory Reporting. https://www.pulitzer.org/winners/staffs-arizona-republic-and-usa-today-network.

Q.ai (7 December 2022). Elon Musk's Neuralink Brain Implant Could Begin Human Trials In 2023. Forbes. https://www.forbes.com/sites/qai/2022/12/07/elon-musks-neuralink-brain-implant-could-begin-human-trials-in-2023/?sh=60ea9f1d147c.

Raftery, Brian (April 01, 2022). How the metaverse and augmented reality will change entertainment. https://ew.com/movies/the-future-metaverse-augmented-reality-entertainment/.

Ravenscroft, Eric. (25 Nov. 2021). "What Is the Metaverse, Exactly?" Wired. https://www.wired.com/story/what-is-the-metaverse/.

Raz, G. (2022). Rage against the empathy machine revisited: The ethics of empathy-related affordances of virtual reality. *Convergence*, 28(5), 1457–1475. https://journals.sagepub.com/doi/10.1177/13548565221086406.

Reader, Ruth (25 Jan. 2022). These women are staking their claim to Web3 and the metaverse. Fast Company. https://apple.news/AlREhXu_IQ3OVeNVSQH-BVw.

Regret Iyer, Shravan, John V. Pavlik and Venus Jin (January 2022). "Virtual Tourism in the Peri-and-Post Covid-19 Era: Understanding How Experiential Media are Utilized in the Making of 'Qatar 2022'." *Advances in Journalism and Communication*. Vol. 10 (2): 81–102. https://www.scirp.org/journal/paperinformation.aspx?paperid=117000.

Reiff, Nathan (4 January 2023). The Collapse of FTX: What Went Wrong with the Crypto Exchange? https://www.investopedia.com/what-went-wrong-with-ftx-6828447.

Reuters (11 April 2022). Meta to Start Testing Money-Making Tools for Its Metaverse. https://money.usnews.com/investing/news/articles/2022-04-11/meta-to-start-testing-money-making-tools-for-its-metaverse.

——— (17 February 2022). Universal Music aims to make money in the Metaverse with NFTs of its artists. https://nypost.com/2022/02/17/universal-music-aims-to-make-money-in-Metaverse-with-artist-nfts/?utm_campaign=iphone_nyp&utm_source=mail_app.

———(21 February 2022). Technology that merges virtual, physical worlds soon to become a reality: Meta. https://www.livemint.com/technology/tech-news/metaverse-technology-that-merges-virtual-physical-worlds-soon-to-become-a-reality-meta-says-11645408692319.html.

—— (26 January 2022). China's Metaverse would be censored, compliant and crypto-less. https://tribune.com.pk/story/2340515/chinas-Metaverse-would-be-censored-compliant-and-crypto-less.

—— (6 July 2021). NFT sales volume surges to $2.5 billion in 2021 first half. https://nypost.com/2021/07/06/nft-sales-volume-surges-to-2-5-billion-in-2021-first-half/.

—— (25 Feb. 2023). Colombia court moves to metaverse, hosts first hearing. https://nypost.com/2023/02/25/colombia-court-moves-to-metaverse-hosts-first-hearing/?utm_campaign=iphone_nyp&utm_source=mail_app.

Richter, Felix (9 Jan. 2020). Smart Speaker Adoption Continues to Rise. https://www.statista.com/chart/16597/smart-speaker-ownership-in-the-united-states/.

Roberts, Daniel (3 March 2023). 'Snow Crash' Author Neal Stephenson Says Future of the Metaverse Won't Require Goggles. Decrypt. https://apple.news/A17YSj-t_S7OZ5O8ZpJswHg.

Robertson, Derek (8 Feb. 2023). How to regulate a universe that doesn't exist. https://www.politico.com/newsletters/digital-future-daily/2023/02/08/how-to-regulate-a-universe-that-doesnt-exist-00081895.

Roettgers, Janko (1 February 2022). "Who owns your address in AR? Probably not you." https://metaverseinsider.tech/2023/04/06/exploring-upland-a-metaverse-journey-through-real-estate-with-idan-zuckerman/.

Rogers, Everett (1994). *A History of Communication Study: A Biological Approach*. NY: The Free Press.

Roose, Kevin (16 February 2023). Help, Bing Won't Stop Declaring Its Love for Me. https://www.nytimes.com/2023/02/16/technology/bing-chatbot-microsoft-chatgpt.html?smid=nytcore-ios-share&referringSource=articleShare.

—— (18 March 2022). What is web3? https://www.nytimes.com/interactive/2022/03/18/technology/web3-definition-internet.html?referringSource=articleShare.

—— (24 March 2021). Buy This Column on the Blockchain! Why can't a journalist join the NFT party, too? https://www.nytimes.com/2021/03/24/technology/nft-column-blockchain.html.

Rosen, Jay. (Winter 1993). Beyond Objectivity. Nieman Reports.

—— (1999). *What are Journalists For?*. Yale University Press. ISBN 9780300089073.

Rosen, Phil (13 Feb. 2023). ChatGPT wrote an article about the market in under a minute. Here's what the buzzy AI is thinking about meme stocks, volatility, and the outlook for 2023. https://markets.businessinsider.com/news/stocks/chatgpt-markets-write-stocks-economy-meme-ai-microsoft-insider-anlayst-2023-1\.

Rosenberg, Louis B. (27 January 2022). "The danger of AI micro-targeting in the Metaverse" https://venturebeat.com/2022/01/27/the-danger-of-ai-micro-targeting-in-the-Metaverse/.

—— (November 19, 2022). No, the metaverse is not dead—it's inevitable. Unanimous A.I. https://venturebeat.com/virtual/no-the-metaverse-is-not-dead-its-inevitable/.

Rosner, Helen (15 July 2021). A Haunting New Documentary About Anthony Bourdain. "Roadrunner," by the Oscar-winning filmmaker Morgan Neville, presents Bourdain as both the hero and the villain of his own story. The New Yorker. https://www.newyorker.com/culture/annals-of-gastronomy/the-haunting-afterlife-of-anthony-bourdain.

Ross Smith, Bayeté (2017) "Firsthand Account: The Assassination of Malcolm X." The Daily 360. https://www.nytimes.com/video/us/100000004817791/malcolm-x-death-new-york-assassination-360.html?smprod=nytcore-iphone&smid=nytcore-iphone-share.

Rubin, Ross (23 Nov. 2022). The metaverse isn't even here yet—but ambient computing is already very real. Fast Company. https://www.fastcompany.com/90814962/the-metaverse-isnt-here-yet-but-ambient-computing-is-already-very-real?partner=rss&utm_source=rss&utm_medium=feed&utm_campaign=rss+fastcompany&utm_content=rss.

Salwen, Michael B.; Garrison, Bruce; Driscoll, Paul D. (2004). Online News and the Public. Routledge. p. 136. ISBN 978-1-135-61679-3.

Sample, Ian (16 June 2022). How Google's chatbot works—and why it isn't sentient. https://apple.news/AblniBo0mRYCPMJQB8-Nn1w.

Sanger, David E. (5 Feb. 2023). Balloon Incident Reveals More Than Spying as Competition with China Intensifies. https://www.nytimes.com/2023/02/05/us/politics/balloon-china-spying-united-states.html?smid=nytcore-ios-share&referringSource=articleShare.

Satariano, Adam and Paul Mozur (7 February 2023). The People Onscreen Are Fake. The Disinformation Is Real. https://www.nytimes.com/2023/02/07/technology/artificial-intelligence-training-deepfake.html.

Sawyers, Paul (28 Feb. 2022). Meta calls for collaborative effort to build the metaverse network infrastructure. VentureBeat. https://venturebeat.com/2022/02/28/meta-calls-for-collaborative-effort-to-build-the-metaverse-network-infrastructure/.

Schlesinger, Arthur Meier (1933). The Rise of the City: 1878–1898 (1933) pp 185–87.

Schwartz, Madeleine, Malika Khurana, Mika Gröndahl and Yuliya Parshina-Kottas (3 March 2023). A Cathedral of Sound. https://www.nytimes.com/interactive/2023/03/03/magazine/notre-dame-cathedral-acoustics-sound.html?smid=nytcore-ios-share&referringSource=articleShare.

Scire, Sarah (16 Nov. 2021). About a third of news organizations have already adopted a remote or hybrid working model. https://www.niemanlab.org/2021/11/about-a-third-of-news-organizations-have-already-adopted-a-remote-or-hybrid-working-model/.

Selinger, Evan, Jules Polonetsky, Omer Tene (2018). The Cambridge Handbook of Consumer Privacy. Cambridge University Press.

Seok, Kwang-Ho, YeolHo Kim, Wookho Son, & Yoon Sang Kim (15 July 2021). Using visual guides to reduce virtual reality sickness in first-person shooter games: Correlation analysis. JMIR Serious Games, 9(3): e18020. doi: 10.2196/18020. PMID: 34264196; PMCID: PMC8323020.

Seymat, Thomas (28 October 2021). Will the Metaverse save Journalism?" https://medium.com/@tseymat/will-the-Metaverse-save-journalism-a4ce044c948e.

Sforza, Lauren (15 Feb. 2023). 78 percent say AI-written news articles would be a bad thing: survey | The Hill. https://thehill.com/homenews/media/3859422-78-percent-say-ai-written-news-articles-would-be-a-bad-thing-survey/.

Shanahan, Paul (2017). The Wall in VR. https://vimeo.com/251397573.

Shead, Sam (13 April 2022). Meta plans to take a nearly 50% cut on NFT sales in its metaverse. https://www.cnbc.com/2022/04/13/meta-plans-to-take-a-nearly-50percent-cut-on-nft-sales-in-its-metaverse.html.

Shearer, Elisa (12 Jan. 2021). More than eight-in-ten Americans get news from digital devices. Pew. https://www.pewresearch.org/fact-tank/2021/01/12/more-than-eight-in-ten-americans-get-news-from-digital-devices/.

Shelley, Mary (1818). Frankenstein; or, the Modern Prometheus.

Sherr, Ian (20 April 2022). Meta Gives First Look at NFL Pro Era Game VR Game Offers People to Play as a Quarterback. https://www.cnet.com/tech/gaming/meta-gives-first-look-at-nfl-pro-era-game-vr-game-offers-people-to-play-as-a-quarterback/.

Shin, D. (2020). User perceptions of algorithmic decisions in the personalized AI system: Perceptual evaluation of fairness, accountability, transparency, and explainability. *Journal of Broadcasting & Electronic Media*, 64(4), 541–565. https://doi.org/10.1080/08 838151.2020.1843357.

Silverstein, Jake (5 Nov. 2015). The Displaced. https://www.nytimes.com/2015/11/08/magazine/the-displaced-introduction.html; https://www.nytimes.com/video/magazine/100000005005806/the-displaced.html.

Sims, Daniel (4 February 2022). "Meta to introduce 'Personal Boundaries' feature in Metaverse." https://www.techspot.com/news/93274-meta-introduce-personal-boundaries-feature-Metaverse.html.

Singh, Jashandeep & Malhotra, Meenakshi & Sharma, Nitin (2022). "Metaverse in Education: An Overview." doi: 10.4018/978-1-6684-6133-4.ch012.

Singh, Sarabjeet (2011). Digital Watermarking Trends. *International Journal of Research in Computer Science* 1. 55–61. doi: 10.7815/ijorcs.11.2011.005.

Singh, Kuldeep (11 April 2020). WebXR—The new web. https://medium.com/xrpractices/webxr-the-new-web-779168b1e6f4.

Skarbez, R., Brooks, Jr., F.P., & Whitton, M.C. (2017). A Survey of Presence and Related Concepts. *ACM Computing Surveys (CSUR)*, 50, 1–39.

Slater, M., Banakou, D., Beacco, A., Gallego, J., Macia-Varela, F., & Oliva, R. (2022). A Separate Reality: An Update on Place Illusion and Plausibility in Virtual Reality. In Frontiers in Virtual Reality (3). https://doi.org/10.3389/frvir.2022.914392).

Slater, M., & Usoh, M. (1993). Representations Systems, Perceptual Position, and Presence in Immersive Virtual Environments. Presence, 2(3): 221–233.

Smith, Christian. VR Is More Dangerous Than Ever, According to Insurance Company. https://www.svg.com/767410/vr-is-more-dangerous-than-ever-according-to-insurance-company/.

Smith, Serena (10 February 2022). "Mental health online: will the Metaverse just make everything worse?" https://www.dazeddigital.com/science-tech/article/55420/1/mental-health-online-internet-social-media-Metaverse-illness-virtual-reality?utm_source=Link&utm_medium=Link&utm_campaign=Flipboard&utm_term=mental-health-online-will-the-Metaverse-just-make-everything-worse.

Somaiya, Ravi (20 Oct. 2015). "The Times Partners with Google on Virtual Reality Project." https://www.nytimes.com/2015/10/21/business/media/the-times-partners-with-google-on-virtual-reality-project.html.

Solomon, Ben C. and Leslye Davis (20 Nov. 2015). "Finding Hope in the Vigils of Paris: A Virtual Reality Film." http://www.nytimes.com/2015/11/21/world/europe/finding-hope-in-the-vigils-of-paris.html?smprod=nytcore-iphone&smid=nytcore-iphone-share&_r=0.

Solove, Daniel J. & Woodrow Hartzog (2022). *Breached! Why Data Security Law Fails and How to Improve it.* Oxford University Press.

Solsman, Joan E. (April 11, 2022). 'The Infinite' VR Is the Closest You'll Feel to Space Without Floating.

Somboon, Thitirat (11 March 2022). "Metaverse: The Future of Education Beyond Frontiers from the Real to the Virtual World." 11 March 2022. https://www.chula.ac.th/en/highlight/67429/.

Soo, Zen (22 Feb. 2023). Baidu to implement ChatGPT-like Ernie Bot chatbot from March. https://apnews.com/article/technology-science-baidu-inc-china-artificial-intelligence-427838070f962108cabed1e553fc3d12.

Spangler, Todd (17 February 2022). iHeartMedia Will Translate English Podcasts Into Other Languages Using Veritone's Synthetic Voices. Variety.com.

Staff of the New York Post (25 March 2008). Headless Body in Topless Bar: The Best Headlines from America's Favorite Newspaper.

Staff Reports, NYTimes (21 Sept. 2022). Rising seas, shrinking coasts: Take an augmented reality tour of damage caused by warming temperatures. https://www.usatoday.com/story/augmented-reality/2022/09/21/interactivestory-experienceid-searise/10391348002/.

—— (13 Aug. 2020). Statues bring suffrage history to you. https://www.usatoday.com/story/augmented-reality/2020/08/13/interactivestory-experienceid-suffragestatues/3358813001/.

—— (16 Nov. 2022). Think you can be a World Cup goalie? Here's your chance to step in their shoes. USA Today. https://www.usatoday.com/story/augmented-reality/2022/11/16/interactivestory-experienceid-soccermen/10671540002/.

—— (7 July 2020). Learn more about destroyed Frederick Douglass statue, in Augmented Reality. USA Today. https://www.usatoday.com/story/augmented-reality/2020/07/07/interactivestory-experienceid-douglass/5390813002/.

Stanley, Alyse (7 April 2022). Lego and Epic Games partnership aims for a kid-friendly metaverse. https://www.washingtonpost.com/video-games/2022/04/07/lego-epic-metaverse-fortnite/.

Statista (9 February 2022). Lifetime sales of Animal Crossing: New Horizons is 37.62 million units of early 2022. https://www.statista.com/statistics/1112631/animal-crossing-new-horizons-sales/.

—— (January 2022). "Global digital population as of January 2021." https://www.statista.com/statistics/617136/digital-population-worldwide/.

Statt, Nick (13 April 2022). Meta's AR ambitions include two pairs of smart glasses. High end and low end in about 2024. https://www.theverge.com/23022611/meta-facebook-nazare-ar-glasses-roadmap-2024.

Steele, Chandra (2 February 2022). "The Metaverse's Big Challenge: Most People Think It's a Privacy Nightmare." https://apple.news/A44qc7SVGToiTGtcD-aPMBQ.

Steen, Jeff (February 2022). "Tim Cook Just Revealed Apple's Plan for Expansion into AR/VR. It's the Opposite of Facebook." https://www.inc.com/jeff-steen/tim-cook-just-revealed-apples-plan-for-expansion-into-ar/vr-its-opposite-of-facebook.html.

Steensen, Jakob Kudsk (2017). Project Tree. MIT. https://www.media.mit.edu/projects/tree/overview/.

Stein, Scott (21 February 2022). Watching Me, Watching You: How Eye Tracking Is Coming to VR and Beyond: Leading eye-tracking company Tobii has some ideas about why this is the next great leap for immersive tech. https://www.cnet.com/tech/computing/watching-me-watching-you-how-eye-tracking-is-coming-to-vr-and-beyond/.

—— (21 March 2022). The Metaverse Isn't a Destination, It's a Metaphor. https://www.cnet.com/tech/computing/features/the-metaverse-isnt-a-destination-its-a-metaphor/.

Stephens, Mitchell (2016). The Voice of America: Lowell Thomas and the Invention of 20th-Century Journalism. St. Martin's Press.

Stephenson, Neal (1992). *Snow Crash*. New York: Bantam Books: 24.

Sternberg, Robert J. "human intelligence". Encyclopedia Britannica, 10 Dec. 2020, https://www.britannica.com/science/human-intelligence-psychology. Accessed 22 February 2022.

Stokel-Walker, Chris (18 January 2023). ChatGPT listed as author on research papers: many scientists disapprove. At least four articles credit the AI tool as a co-author, as publishers scramble to regulate its use. https://www.nature.com/articles/d41586-023-00107-z.

Summerson, Cameron (11 April 2018). What Are the ARCore and ARKit Augmented Reality Frameworks? https://www.howtogeek.com/348445/what-are-the-arcore-and-arkit-augmented-reality-frameworks/.

Sullivan, Mark (21 February 2022). What are decentralized autonomous organizations—and why should you care? Fast Company: https://apple.news/AFRt2KM8bR0OSLGP3KtzF7A.

—— (5 March 2022). How Epic Games is changing gaming—and maybe the metaverse. Fast Company: https://apple.news/AFqANmo65SkCcl8dUyW6jKA.

Sundar, S. Shyam (2017). Being There in the Midst of the Story. https://www.academia.edu/71883193/Being_There_in_the_Midst_of_the_Story_How_Immersive_Journalism_Affects_Our_Perceptions_and_Cognitions.

Sundar, S. Shyam, Jin Kang and Danielle Oprean (Nov. 2017). "Immersive Journalism Affects our Perceptions and Cognitions." *Cyberpsychology, Behavior, and Social Networking.* 20 (11): 672. doi: 10.1089/cyber.2017.0271.

Sunstein, Cass (2015). "The Ethics of Nudging." *Yale Journal on Regulation*. Issue 2.

Swayne, Matt (28 February 2022). "Top 5 Universities for Metaverse Research." https://metaverseinsider.tech/2022/02/28/top-5-universities-for-metaverse-research/.

Tabuchi, Hiroko (13 April 2022). NFTs Are Shaking Up the Art World. They May Be Warming the Planet, Too. https://www.nytimes.com/2021/04/13/climate/nft-climate-change.html?referringSource=articleShare.

Takahashi, Dean (4 March 2022). The DeanBeat: Our Metaverse Forum takes a shot at defining the metaverse. https://venturebeat.com/2022/03/04/the-deanbeat-our-metaverse-forum-takes-a-shot-at-defining-the-metaverse/.

—— (Jan 4, 2023). MeetKai has rolled out new creators tools as part of its ambition to create a portfolio of metaverse and conversational AI technologies. VentureBeat: https://apple.news/AATc4E31UTqar91UY8a0V1A.

TAL Technologies (2022). "A brief overview of TCP/IP communications." https://www.taltech.com/datacollection/articles/a_brief_overview_of_tcp_ip_communications.

Tangcay, Jazz (15 July 2021). Anthony Bourdain's AI-Faked Voice in New Documentary Sparks Backlash. https://www.msn.com/en-us/movies/news/anthony-bourdain-s-ai-faked-voice-in-new-documentary-sparks-backlash/ar-AAMcxLz.

Tashev, Ivan J. (2019). Capture, Representation, and Rendering of 3D Audio for Virtual and Augmented Reality. *International Journal on Information Technologies & Security*, № SP2 (vol. 11). https://ijits-bg.com/sites/default/files/archive/2019%28vol.11%29/contents-vol11-2019.pdf.

Tay, Daniel Thomas (20 Feb. 2023). Metaverse live streaming is more than just watching shows in a 3D virtual world. https://venturebeat.com/virtual/metaverse-live-streaming-is-more-than-just-watching-shows-in-a-3d-virtual-world/.

Technavio (22 July 2022). Metaverse Real Estate Market Size to grow by USD 5.37 billion, Market Driven by Growing Popularity of Mixed Reality & Cryptocurrency—Technavio.

https://www.prnewswire.com/news-releases/metaverse-real-estate-market-size-to-grow-by-usd-5-37-billion-market-driven-by-growing-popularity-of-mixed-reality--cryptocurrency---technavio-301591153.html.

The Guardian (8 September 2020). "A robot wrote this entire article. Are you scared yet, human? GPT-3. https://www.theguardian.com/commentisfree/2020/sep/08/robot-wrote-this-article-gpt-3.

TheNationThailand (2 March 2022). Thammasat heads into virtual reality world with metaverse campus. https://www.nationthailand.com/in-focus/40012958.

Thompson, Clive (April 14, 2016). The Minecraft Generation. NYTImes.com. https://www.nytimes.com/2016/04/17/magazine/the-minecraft-generation.html.

Thorbecke, Catherine (February 13, 2021). GameStop timeline: A closer look at the saga that upended Wall Street. https://abcnews.go.com/Business/gamestop-timeline-closer-saga-upended-wall-street/story?id=75617315).

Tilley, Aaron (3 March 2023). Apple Approves ChatGPT-Powered App After Assurance of Content Moderation. https://www.wsj.com/articles/apple-approves-chatgpt-powered-app-after-assurance-of-content-moderation-9c82cd7.

Tiku, Nitasha (October 7, 2022). "'Chat' with Musk, Trump or Xi: Ex-Googlers want to give the public AI". *The Washington Post*. https://www.washingtonpost.com/technology/2022/10/07/characterai-google-lamda/.

Tilleke & Gibbins. January 31 2022. Immersing Intellectual Property Rights in the Metaverse. https://www.lexology.com/library/detail.aspx?g=f88ba949-242d-4295-98c6-2d73415d2709.

TIMEImmersive (18 July 2019). Welcome to TIME Immersive's Apollo 11 'Landing on the Moon' Experience. https://time.com/longform/apollo-11-moon-landing-immersive-experience/.

Times R&D (16 June 2022). Spatial Journalism: Showcasing 31 Published Experiments in AR Storytelling. https://rd.nytimes.com/projects/augmented-reality-storytelling.

—— (16 June 2022). Bronx Fire. https://rd.nytimes.com/projects/augmented-reality-storytelling#bronx-fire.

—— (12 August 2022). California Megastorm. https://rd.nytimes.com/projects/augmented-reality-storytelling#california-megastorm.

—— (28 Oct. 2021). Wildfire Storms. https://rd.nytimes.com/projects/augmented-reality-storytelling#wildfire-storms.

Tong, Goh Chiew (15 Nov. 2022). 'No longer science fiction'? Metaverse could pump $1.4 trillion a year into Asia's GDP, report says. https://www.cnbc.com/2022/11/14/metaverse-could-pump-1point4-trillion-a-year-into-asias-gdp-deloitte.html.

Torkington, Simon (23 March 2022). 4 things you need to know about the metaverse this week. https://www.weforum.org/agenda/2022/03/metaverse-stories-this-week-23-march/.

Tracy, Phillip (18 February 2022). Ugh, Zuckerberg's Metaverse Isn't as Dead as We All Hoped Mark Zuckerberg and his "Metamates" aren't the only ones roaming around the Metaverse. Gizmodo: https://apple.news/ADvDsirs5SoynOujSMfOAPg.

Tran, Kevin (14 February 2022). "Survey: How Metaverse Meets Media in the Minds of Consumers. Variety.com. https://variety.com/vip/survey-how-Metaverse-meets-media-in-the-minds-of-consumers-1235178886/.

Turing, Alan (October 1950), "Computing Machinery and Intelligence", *Mind*, LIX (236): 433–460, doi: 10.1093/mind/LIX.236.433, ISSN 0026-4423.

Ultimate History of CGI (5 February 2018). "Computer Animated Hand (1972)—First polygonal 3D animation. https://www.youtube.com/watch?v=fAhyBfLFyNA.

USAToday (16 Nov. 2022). Think you can be a World Cup goalie? Here's your chance to step in their shoes. https://www.usatoday.com/story/augmented-reality/2022/11/16/interactivestory-experienceid-soccermen/10671540002/.

—— (22 June 2021). GANNETT Launches Its Inaugural Non-Fungible Token Inspired by the First Newspaper on the Moon. https://www.usatoday.com/story/news/pr/2021/06/22/gannett-launches-its-inaugural-non-fungible-token-inspired-first-newspaper-on-the-moon/5311544001/.

—— (7 July 2020). Learn more about destroyed Frederick Douglass statue, in Augmented Reality. https://www.usatoday.com/story/augmented-reality/2020/07/07/interactivestory-experienceid-douglass/5390813002/.

—— (7 May 2020). No mask? Here's how far your germs can travel. https://www.usatoday.com/story/augmented-reality/2020/05/07/interactivestory-experienceid-aerosol/3047773001/.

Vettehen, Paul Hendriks, Dann Wiltink, Maite Huiskamp, Gabi Schaap and Paul Ketelaar (2019). "Taking the full view: How viewers respond to 360-degree video news." https://www.sciencedirect.com/science/article/abs/pii/S0747563218304588.

Vincent, James (5 Jan. 2023). Top AI conference bans use of ChatGPT and AI language tools to write academic papers. Are AI writing tools just assistants or something more? The Verge: https://apple.news/AcxshFQ9TSOmN-OklurUXAw.

—— (8 Feb. 2023). Google's AI chatbot Bard makes factual error in first demo. https://www.theverge.com/2023/2/8/23590864/google-ai-chatbot-bard-mistake-error-exoplanet-demo.

Virbela (13 January 2022). Three Universities Already Teaching Classes in the Metaverse. https://www.virbela.com/blog/three-universities-already-teaching-classes-in-the-metaverse.

Virgilio (9 February 2022). "What Comparisons Between Second Life and the Metaverse Miss." Slate.com https://slate.com/technology/2022/02/second-life-metaverse-facebook-comparisons.html.

Virtual Reality Society (2018). "History of Virtual Reality." http://www.vrs.org.uk/virtual-reality/who-coined-the-term.html.

Vivarellii, Nick (1 April 2022). Dubai-Based MContent Partners plans to fund 100 metaverse content productions in 2022. https://variety.com/2022/digital/news/mcontent-pwc-ripple-vs-sec-saga-metaverse-1235220838/.

Wall Street Journal, The (29 Nov. 2015). Japan Is Changing How We'll Grow Old. http://www.wsj.com/articles/virtual-reality-video-japan-is-changing-how-well-grow-old-1448809232.

—— (9 April 2022). The Future of Socializing at Work? Virtual Golf. https://www.wsj.com/articles/the-future-of-socializing-at-work-virtual-golf-11649476854.

WAN-IFRA (21 March 2022). Automated content fills in reporting gaps at US local media group McClatchy.

—— (31 May 2023). https://wan-ifra.org/2023/05/new-genai-survey/.

Wang, Brian (24 Feb. 2023). What to Expect for OpenAI GPT-4 and GPT-5. https://www.nextbigfuture.com/2023/02/what-to-expect-for-openai-gpt-4-and-gpt-5.html.

Wang, Zac (22 March 2022). See the pitch deck an AR startup used to raise $13 million in seed funding to bring the metaverse to life. https://www.businessinsider.com/pitch-deck-ar-startup-auki-labs-raises-13-million-2022-3.

Warren, Rachel, Demitri Kalogeropoulos and Jose Najarro (11 Feb. 2022). These Companies Are Winning the Metaverse Race. https://www.fool.com/investing/2022/02/11/these-companies-are-winning-the-Metaverse-race/?source=isafpbcs0000001&utm_source=flipboard&utm_medium=feed&utm_campaign=firehose.

Warrick, Ambar (9 Jan. 2023). Microsoft plans $10 billion investment in ChatGPT owner OpenAI-Report. https://www.investing.com/news/stock-market-news/microsoft-plans-10-billion-investment-in-chatgpt-owner-openai-report-2976695.

Washington Post Staff (20 Feb. 2023). The Post's coverage of Amazon rainforest destruction wins a Polk Award. https://www.washingtonpost.com/media/2023/02/20/polk-award-terrence-mccoy-amazon-undone/.

—— (17 Feb. 2023). The new Bing told our reporter it 'can feel or think things'. WashingtonPost.com https://www.washingtonpost.com/technology/2023/02/16/microsoft-bing-ai-chat-interview/.

—— (16 March 2022). NFTs and Play-to-Earn Games Shed Light on Future of Metaverse—The Wall Street Journal. https://www.wsj.com/video/series/in-depth-features/nfts-and-play-to-earn-games-shed-light-on-future-of-metaverse-economy/A5540926-5121-48CC-9248-78478BEAF38D?reflink=article_email_share&st=cgrwmizfr4e8cn4.

WashPostPR (10 May 2016). The Washington Post releases augmented reality view of Freddie Gray's case. https://www.washingtonpost.com/pr/wp/2016/05/10/the-washington-post-releases-augmented-reality-view-of-freddie-grays-case/.

Watson, Amy (2 June 2022). News in the United States—statistics & facts. Statista.com. https://www.statista.com/topics/1640/news/; https://www.statista.com/topics/994/newspapers/#topicOverview; https://www.statista.com/statistics/242659/revenue-of-the-us-newspaper-publishing-industry/.

Wayt, Theo (22 December 2021). Elon Musk says Metaverse isn't 'compelling,' Web3 mostly 'marketing." https://nypost.com/2021/12/22/elon-musk-says-idea-of-Metaverse-is-not-compelling/?utm_campaign=iphone_nyp&utm_source=mail_app.

Werbach, Kevin (2018). The Blockchain and the New Architecture of Trust. MIT Press.

White, Monica J. (1 Febuary 2022). Woman Says She Was 'Virtually Gang-Raped' in Facebook's Metaverse. https://www.vice.com/en/article/3abpg3/woman-says-she-was-virtually-gang-raped-in-facebooks-Metaverse.

Whyte, Kenneth (2009). The Uncrowned King: The Sensational Rise of William Randolph Hearst. Berkeley: Counterpoint. ISBN 978-1582439853.

Wi-Fi 6 will soon go mainstream and usher in the Metaverse https://www.digitaltrends.com/computing/wi-fi-6-may-hit-60-percent-market-share-in-2022-becoming-mainstream/.

Wijnberg, Rob (7 Oct. 2017). Why objective journalism is a misleading and dangerous illusion. https://thecorrespondent.com/6138/why-objective-journalism-is-a-misleading-and-dangerous-illusion/157316940-eb6c348e.

Wikipedia (25 January 2022). "Greta Thunberg." https://en.wikipedia.org/wiki/Greta_Thunberg.

Wile, Rob (19 December 2022). Fortnite maker Epic Games fined $520M after accusations it exposed young players to potential harm. NBCNews.com https://www.nbcnews.com/business/consumer/fortnite-maker-epic-games-fined-520-million-accusations-exposed-child-rcna62369.

Williams, James (2023). Virtual Reality Demographics: 57 User Facts & Numbers. Techpenny https://techpenny.com/virtual-reality-demographics-57-user-facts-numbers-2023/.

Winslow, George (19 July 2021). New Data on 'The State of TV News'. TVTech.com. https://www.tvtechnology.com/news/new-data-on-the-state-of-tv-news.
Witt, Toni (19 March 2022). XR pioneer calls for metaverse regulation. https://venturebeat.com/2022/03/19/xr-pioneer-calls-for-metaverse-regulation/.
WooGlobe (2023). 'Extended Reality enthusiast plays 3D Chess on HoloLens2 *The Future is Now*' www.youtube.com/watch?v=W6FJd8pgdA4.
Wu, H., Cai, T., Luo, D., Liu, Y., & Zhang, Z. (2021). Immersive virtual reality news: A study of user experience and media effects. *International Journal of Human-Computer Studies*, 147. https://doi.org/10.1016/j.ijhcs.2020.102576.
Wurmser, Yoram (13 April 2022). AR and VR enter the mainstream. https://www.emarketer.com/content/ar-vr-enter-mainstream.
Yahoo!finance (19 January 2023a). Davos 2023: Future of the Metaverse. https://finance.yahoo.com/video/davos-2023-future-metaverse-152311205.html.
—— (27 Jan. 2023b). Meta stock continues rebound in 2023 as tech companies pivot further into A.I., VR. https://finance.yahoo.com/video/meta-stock-continues-rebound-2023-203200064.html?guccounter=1&guce_referrer=aHR0cHM6Ly9kWNrZHVja2dvLmNvbS8&guce_referrer_sig=AQAAAFoaVt0Jgmrz33m4F8LQ93tU37YKNp45BNiRU295yW5YSO9PkirImoDOfHuNU3b9C9E4zFl6W1R6mSZrM5SFK7y0Au9ht-wMKUfBuhQ0Y254BzjvjaYypRR60aUlPgL3xfTqaM81r3SoP4-QEwJDllW3avB5uVyxoyu0Src7ffpU.
Yanev, Victor (4 January 2022) "Video Game Demographics—Who Plays Games in 2021." https://techjury.net/blog/video-game-demographics/.
Young, Jin You and Matt Stevens (29 Jan. 2023). Will the Metaverse Be Entertaining? Ask South Korea. https://www.nytimes.com/2023/01/29/business/metaverse-k-pop-south-korea.html?smid=nytcore-ios-share&referringSource=articleShare Will the Metaverse Be Entertaining? Ask South Korea.
Zahn, Max (11 Nov. 2022). A timeline of Elon Musk's tumultuous Twitter acquisition. https://abcnews.go.com/Business/timeline-elon-musks-tumultuous-twitter-acquisition-attempt/story?id=86611191.
Zaiets, K., Thorson, M., Sullivan, S. J. & Haseman, J. (2020, Nov 13). US coronavirus map: Tracking the outbreak: Track coronavirus outbreaks across the US and in your state with daily updated maps, total cases and deaths. USA Today. https://www.usatoday.com/in-depth/graphics/2020/03/10/us-coronavirus-map-tracking-united-states-outbreak/4945223002/.
Zeitchik, Steven (29 March 2022). Hackers hit popular video game, stealing more than $600 million in cryptocurrency. https://www.washingtonpost.com/technology/2022/03/29/axie-infinity-cryptocurrency-hack/.
—— (4 February 2022). "This is what it feels like to attend a film festival in the Metaverse." https://www.washingtonpost.com/technology/2022/02/04/sundance-metaverse-quest-virtual-reality/.
Zhu, H. (2022b). *MetaAID: A Flexible Framework for Developing Metaverse Applications via AI Technology and Human Editing.* arXiv preprint arXiv:2204.01614.

INDEX

360-degree 4, 6, 11, 12, 14, 15, 30, 65, 113, 114, 117, 120–22

accuracy 14, 25, 30, 75, 76, 78, 80, 86, 99, 103, 105, 132
ADVENT (*Colossal Cave Adventure*) 133
advertising-supported video on demand (AVOD) 18
affordances 14, 39, 57–60, 64, 78, 79, 97, 112, 131, 135, 140
africarare 50
After Solitary 120
algorithms 5, 68, 69, 77, 78, 80–83, 90, 130, 136
AltSpaceVR 48
Amazon Echo with Alexa 72, 104, 108
ambient computing 54
Apple 32, 52, 60, 72, 104, 105, 108, 123
Apple Homepod ARKit 72, 104
Arizona Republic, The 121
artificial intelligence (AI)
  Artificial General Intelligence (AGI) 91–94
  Artificial Narrow Intelligence (ANI) 91–93
  Artificial Super-Intelligence (ASI) 91–94
  generative AI 18, 26, 27, 31, 42, 49, 66, 68–70, 80, 84–89, 92, 100, 102, 103–7, 109
  tiers 90–93
augmented reality (AR) 1, 4, 11, 17, 20, 30, 31, 41, 43, 47, 50–55, 57–61, 65–68, 70, 94, 97, 99, 100, 110–15, 122–30, 134
  persistent 124

Automated Insights Wordsmith 82, 91
Axie Infinity 47, 55
Baidu Ernie 88
Ball, LaMelo 56
BBC *Civilisations AR* 125, 126
Berners-Lee, Sir Tim 53
bias 25, 27, 37, 59, 75, 87, 105, 107, 131–33, 136
Bitcoin 44, 54, 96
Black Metaverse 50
blockchain Ethereum 50, 54
Boudinot, Elias 142
brain-computer interface (BCI) 5, 68
buzzfeed 102
ByteDance 60, 73

capital 52, 53
CBS News 90
Cerf, Vinton 16, 39
CGI animation 18, 93
character.ai 100–2
cloud 69, 73, 74, 82
CNET 86
CNN 8, 55, 80, 114
Columbia Newsblaster 80
cookies 33
Cosmopolitan 105
COVID-19 3, 10, 16, 20, 26, 57, 98, 129, 140
Craigslist 41, 43
creativity 10, 22, 89–90
creator economy 19, 50
critical cultural 135

crossplay 51
cryptocurrency 17, 21, 26, 39, 44, 55, 89, 92, 114, 133, 136

data collection 33, 36, 64, 80
data mining 110
datafication 35
de la Peña, Nonny 48, 121
Decentraland 47
decentralized autonomous organization (DAO) 54
deepfake 32, 41, 91, 105, 107, 109
degrees of freedom (DoF) 4, 114, 119–22, 129
Deloitte 61
Des Moines Register 117, 125
Diallo, Amadou 30, 115
digital estate 23
Disney 18
disruption 1, 16, 21, 30, 32, 53
doorway effect, the 138
dynamic immersive news experience (DINE) 76, 96, 132

economics, of news 112
editorial independence 25–28, 36, 131
Emblematic Group 120
embodiment 13, 59, 138, 140
empathy 11–13, 21, 57, 59, 98, 99, 116, 121, 138, 140
encryption 73
engagement 4–6, 8, 10, 12, 13, 17, 18, 23, 36, 54, 57–59, 64, 72–74, 76, 93, 99, 112–14, 133–36, 140, 141
Epic Games 9, 10, 51
Epik Prime 50
Ertha 50
ethical practice 10, 25, 28–29, 36, 78, 131
experiential media (EM) theory 112–13, 117, 118, 130, 131
eXtended Reality (XR) 4, 11, 20, 31, 41, 43
Eye-tracking 99, 100

facial recognition 74, 75
fact-checking 105

field of dreams model 17
first-person perspective
Fortnite 9, 10, 13, 46
Fox News 8, 80
freedom of speech and press 32, 39, 102, 132, 135

geographic information systems (GIS) 107
Github 113, 129
Google
  ARCore 123
  Bard 86, 87
  Home 72
  YouTube 48, 55, 82, 114, 115
Guardian, The 84

haptic 5, 14, 15, 18, 19, 22, 46, 51–53, 59, 64, 68, 75, 99, 111, 113, 120, 121, 126, 140, 141
Harvest of Change 117
head-mounted display (HMD) 17–19, 46, 48, 57, 60, 73, 113, 114, 119, 136, 137
Heilig, Martin 57
historically Black colleges and universities (HBCUs) 50
holography 52
  tensor 19
Home After War: Returning to Fear in Fallujah 122
Hunger in LA 121

IBM's Watson 83, 91
identity 22, 30, 31, 71, 78, 109, 121, 138, 139
immersive education 97–99
intellectual property (IP) 28, 35, 40, 54, 56–57, 86, 134
interactive 3–6, 10, 11, 15, 17–19, 21, 22, 30, 35, 41, 43, 44, 47, 52, 57, 65–67, 72, 76, 78–80, 94, 96, 100, 105, 111, 121, 123–30, 134, 141
Internet of Things (IoT) 54, 69, 93

Journalism principles 25–37
JPMorgan Chase 20

Kahn, Robert 16, 39
King Sejong Institute 47
Klaytn 47

LA Times 69, 81, 82
Lanier, Jaron 22, 140
Lasswell, Harold 27
Lazarsfeld, Paul 141
light detection and ranging (LiDAR)
  66, 70

machine learning (ML) 69, 70, 72, 100
machine vision (MV) 69, 74
MeetKai 74
mental health 14, 31, 57, 58, 110
Mergerson, Cristoph 31
Merton, Robert K. 16
Meta
  Facebook 19, 32, 35, 44, 114
  Horizon Worlds 19, 31, 46, 47, 58, 59, 114
  Instagram 35, 74, 125, 127
  Oculus 60, 116
  Quest 13, 20, 31, 46, 59, 60, 73, 114,
    116, 122
Metacity 50
Microsoft
  Bing 103
  Cortana 104
  Hololens 52
  Minecraft 46, 48, 49, 114
  Sydney 103, 104
  VALL-E 109
  Xbox 31, 47, 58
Mirandus 47
misinformation 31, 41, 54, 73,
  133, 140
mixed reality (MR) 4, 11, 17, 52, 57,
  59, 60, 66, 87, 105, 111, 113, 114,
  124, 140
Moore's Law 19
MTN 50
multisensory 5–6, 57, 59, 63, 64, 66, 76,
  85, 96, 104, 111, 112, 130,
  136, 139
Multiplicative Power of Masks, The 129
Murrow, Edward R. 141
Musk, Elon 5, 16, 40, 41, 84, 93, 100

Narrative Science 91
  Quill 82
natural language processing (NLP) 70,
  80, 81, 86
natural user interface (NUI) 82, 93, 112,
  113, 131
NBC News 8, 9, 117
NBD AI TV 88, 89
Netflix 33, 53
neural network 66, 69, 85, 86, 100,
  102, 106
Neural Radiance Field (NeRF) 66
Neuralink 5, 93
New York Post, The 90
New York Times
  The Displaced 116, 119
  virtual crime scene 128
  Wildfire Storms in AR 126
news avatar 72, 77–80, 84, 132
  three laws 78
news council, virtual 132, 133
Nintendo Switch *Animal Crossing:*
    New Horizons 10
nonfungible tokens (NFTs)
  21, 44
Notre-Dame de Paris 67, 68
NowHere Media 122

objectivity 113
Onyx Lounge 47
OpenAI
  ChatGPT 79–81, 86–88, 91, 92, 100–3,
    105, 107–9
  ChatGPT Plus 87, 92
  DALL-E 86, 105–7
  GPT-3 84–87
  GPT-4 87, 88

Pantheon of Gay Mythology 94
PBS Frontline 120
photogrammetry 65, 120, 128
platform 1, 4, 6–10, 13, 17, 18, 20, 21,
    26–28, 31–34, 36, 37, 39–54, 56–66,
    70, 72, 73, 75, 76, 78, 79, 81–83, 86,
    88, 89, 93, 94, 97, 98, 100, 102, 103,
    105–12, 114, 116, 117, 121, 122, 125,
    129, 131, 132, 134, 135, 139

Pokémon Go 54
post-traumatic stress syndrome
    (PTSD) 31
pre-Metaverse 6, 9, 10, 21
presence 1, 11–14, 16, 21, 31, 36, 41,
    44, 46, 49, 50, 54, 56, 59, 85,
    102, 114, 120, 131, 132, 135, 136,
    138, 140
privacy 10, 22, 28, 30–35, 39, 40, 41, 54,
    58, 74, 95, 100, 102, 103, 109, 110,
    135, 136
propaganda 25, 27, 88, 89, 105
Proteus effect, the 138
psychological impact 63
PwC 98, 99

Quakebot 81, 82, 91, 94
Quartz 124
Quest 13, 20, 31, 46, 59, 60, 73, 114, 116,
    120, 122

regulation, regulatory 27, 33, 41, 74,
    107, 110
relational considerations 139
Reuters 67, 80, 82
Roblox 49
robotics, robots virtual 6, 26, 70, 72, 78,
    92, 133
Roku 20

Sandbox 31, 47, 58
Satire, Parody 40
Second Life 6, 7, 9, 13, 44, 46, 79, 114
security 5, 20, 33, 34, 103
Seeger, Pete 29, 30
Sensorama 57
sexual assault, harassment 30, 31,
    65, 139
Shelley, Mary 17
singularity, the 5
smart media 1, 70, 72–76, 93, 94
    smartphone 1, 19, 20, 70–74, 92, 99,
        117, 119, 123–25, 128
Snow Crash 1
Sony PS VR2 52
spatial computing 4
    audio/sound 14

Stephenson, Neal 1, 2, 18, 133
Stereoscope 57
streaming media 18, 53, 61, 96
structural and systemic theory 134
subscription 8, 18, 27, 41, 42, 87
surveillance capitalism 22, 31, 135
Sutherland, Ivan 57
Sword of Damocles 57

TechMVerse 50
Thomas, Lowell 141
three dimensional (3D) 4, 46
    scene reconstruction 65, 66, 80
TikTok 75
TimeImmersive 124
trademarks 35, 56
Transmission Control Protocol (TCP) and
    Internet Protocol (IP) (TCP-IP) 16
transparency 25, 35–36, 56, 105, 129,
    131, 133
trust 26, 28, 34, 35, 36, 37, 74, 75, 81, 87,
    101, 104, 105, 133, 136, 138
truth 23, 25, 30, 36, 37, 40, 44, 47,
    76, 78, 80, 89, 90, 94, 102, 131,
    133, 142
Turing Test 85
Turing, Alan 104
Twitter 127

unintended consequences 16–17,
    31–32, 121
USA Today Network 55, 121, 124–26
usage user
    user experience (UX) 5–6, 11, 14, 18,
        52, 57, 59, 96, 97, 99, 100, 120, 135,
        137, 138, 140
    user tracking 34, 36, 58

videogames 18
    simulation 13, 44, 87, 99, 105, 133
Virbela 48
virtual reality (VR)
    cinematic VR 12, 14, 15, 21, 26, 84,
        94, 97
    VR sickness addiction 97, 137
virtual voice, synthetic 109
Virtualand 47

Vive Viverse 46, 47
volumetric 19, 66, 12
VRChat 21

The Wall 121, 122
Wang, Cher 46
Washington Post, The 123, 124
    Sport Climbing 128

watermark, digital 105, 106
We Met in VR 66, 67, 139
Web3 48, 60, 61, 127, 132
WebXR journalism 127–29, 132

XiRang 46

Zuckerberg, Mark 16, 51

Milton Keynes UK
Ingram Content Group UK Ltd.
UKHW021918220724
445762UK00002B/11